PHYSIOLOGICAL VARIATION
AND ITS GENETIC BASIS

SYMPOSIA OF THE
SOCIETY FOR THE STUDY OF HUMAN BIOLOGY

Volume XVII

PHYSIOLOGICAL VARIATION AND ITS GENETIC BASIS

Edited by
J. S. WEINER

TAYLOR & FRANCIS LTD

LONDON

HALSTED PRESS

(a division of John Wiley & Sons Inc.)

NEW YORK-TORONTO

1977

First published 1977 by Taylor & Francis Ltd, London and
Halsted Press (a division of John Wiley & Sons Inc.), New York

© 1977 Taylor & Francis Ltd

Printed and bound in Great Britain by Taylor & Francis (Printers) Ltd, Rankine
Road, Basingstoke, Hampshire.

Taylor & Francis ISBN 0 85066 108 0

Library of Congress Cataloging in Publication Data

Main entry under title:
Physiological variation and its genetic basis.
    (Symposia of the Society for the Study of Human Biology; v. 17)
    Included indexes.
    1. Human genetics—Congresses. 2. Human physiology—Congresses. 3.
Variation (Biology)—Congresses. I. Weiner, Joseph Sidney, 1915- II. Series:
Society for the Study of Human Biology. Symposia; v. 17. [DNLM: 1. Variation
—Congresses. 2. Genetics, Population—Congresses.
W3 S05915 v. 17 1976/QH401 P578 1976]
QH431.P52 573.2'2'1 77-14159
ISBN 0-470-99314-6

CONTENTS

v

Contents

INTRODUCTION

THE EDITOR of this volume, and organizer of the symposium on which it is based, is well aware that the enterprise represents an excursion into difficult, largely unknown and even dangerous territory. Nevertheless, so fundamental are physiological responses and attributes for the efficient and adaptive functioning of the human organism that the need to understand the causes, sources and limits of their variation, both temporal and populational, presses heavily on the human biologist. None of the contributors to this topic is so naïve as to assert dogmatically that for the polygenic systems that are believed to underly the majority, if not all, physiological characteristics, the observed variance can be separated into fixed genetic and non-genetic components. As the papers show, it is the essence of these characters that they are highly responsive to environmental change, in both the short and long term. The variable intensity of expression of these characters will of necessity be reflected in a changing relationship between acquired and in-born factors in determining differences in response between individuals within a group subject to a range of environmental exposure. This, in fact, constitutes the central problem in the analysis of physiological variability and is the major theme running through this volume.

The physiologically minded biologist who goes into the field of population, ethnic or ecological study as was done extensively in the Human Adaptability Section of the International Biological Programme (see P. T. Baker and J. S. Weiner (eds), 1966, *The Biology of Human Adaptability*, Oxford University Press; K. J. Collins and J. S. Weiner, 1977, *Human Adaptability: A History and Compendium of Research*, Taylor & Francis) will have his own approach to the variability he encounters. He may see it as related adaptively to the requirements of the habitat; he may consider that age, sex, or body size or composition are determining or associated factors; he may discover that long-term exposure exerts a strong influence; but in the final analysis he is bound to ask how far, if at all, inherited differences account for the inter-individual or the intra-group variability. The problem this poses is whether his field design has been adequate to allow any insight into these questions. Many, quite often striking, observed population contrasts leave the question of a genetic contribution unclear or ambiguous.

To identify a genetic component in characters which show continuous quantitative variation, the extent of which may differ in different environments, or in different communities, poses a major challenge for the human geneticist. This has far-reaching implications. It enters into out understanding of variation in development, growth and the ageing process; it has become a highly controversial issue in the behavioural and psychological domain; but its application to physiological functioning is still at an early stage. It was the aim of the symposium to encourage a more deliberate and more intensive interest in the analysis of phenotypic physiological variation.

The first three papers of this volume present the principles and concepts as well as the methods which are involved in the genetic analysis of continuous variation. Gibson examines the biometrical approach to the partition of variation using data from relatives of varying degree, including sibships and twins. He advocates the use of genetic markers to identify polygenic control of variation and explains the detection of linkage between a marker locus and a quantitative trait using familial data. He also urges the search within geographically distinct populations for associations between marker genes and functional traits.

These issues are taken up by Edwards who urges caution in the interpretation of phenotypic correlations and points to difficulties in detecting the presence of major alleles through twin or family studies. He too emphasizes the value of looking for associations between markers and measurable traits. These may well indicate the action of a few strong alleles rather than of a large number of loci with small effects. Like Gibson, he points out the need for large samples of individuals or families.

Roberts extends the discussion of the approaches presented by Gibson and Edwards and illustrates, from his recent work, the methods available for seeking for genetic factors in physiological variation. These range from simple heritability estimations from first degree relatives to extended family studies and to intrafamilial analysis within whole populations. The value of marker genes within polygenes determining physiological variation is well demonstrated.

His discussion of the effects of environmental change on genotypic distribution is very relevant to the populational and epidemiological analysis presented by a number of contributors to this volume.

In the next four contributions the interaction between heredity and environment is examined intensively by means of twin studies. These papers are concerned with a topic of major physiological significance—the components of the oxygen transport system. Klissouras finds that the variance in the maximal O_2 uptake during exercise is largely genetic in origin in individuals who have lived under similar environmental conditions; and that this is true also for anaerobic capacity and maximal muscle force; but the genetic component is small in most lung responses (e.g. forced expiratory volume, total lung volume, residual volume). These two conclusions gain support from the results presented by Leitch on athletic performance and by Cotes and his colleagues on respiratory function. In contrast to Klissouras's findings on oxygen transport during muscle work, the variance of muscle structure itself and its enzymatic content, according to Howald's twin study, reflects the influence of environmental factors and habitual activity much more than that of genetic factors.

The existence of a major genetic determinant in the variance of blood pressure is strongly supported, in Miall's contribution, by twin and family analysis. As he shows, the most convincing explanation lies in a polygenic hypothesis. Paradoxically, the character with which Annett is concerned, handedness, seems at first sight easily classifiable and analysable into discrete classes, but turns out to be based on a continuous distribution (with a dextral bias) reflecting graded differences between the hands in skill. Unlike blood pressure variability, the distribution of handedness can be accounted for (in line with arguments advanced by Gibson and Edwards) by the operation of a few alleles or genetic factors. Annett's model is also a good example of applying family analyses to a population since her theory gains support from the distribution of handedness amongst dysphasics.

The material presented by Weiner and by Knip on sweat gland activity is not based on family or twin investigations but is representative of the "anthropological" or "ethnic" approach which has been widely used in an attempt to discover whether population genetic differences have any influence on determining the extent of short-term or long-term response or adaptation to a particular stress (e.g. climate, nutritional, pathogenic). The authors do not find convincing evidence for a major genetic influence in determining inter or intrapopulational differences in the number of active sweat glands, or the intensity of sweat gland response to heat stress. The observed variability (apart from factors of age and sex), is very largely acquired through differing degrees of heat exposure.

It is by no means easy (or perhaps advisable) to partition these simple population comparisons into genetic and non-genetic components. An effective design for such investigations is that proposed by Harrison (in Baker and Weiner, *op. cit.*) and by Baker (in *The Measure of Man*, ed. E. Giles and J. S. Friedlander, 1976, Cambridge, Mass.). This requires a 2×2 matrix, using two populations of demonstrably different genetic constitution and both exposed to living in two contrasting situations. Weiner, for example, quotes findings on African (Bantu) and European groups before and after acclimatization to heat. This design is formally analogous to the "DZ and MZ twins reared together and apart" design.

This strategy is adopted by Schull and Rothhammer in their comprehensive study aimed to appraise the genetic contribution to man's adaptation (anatomical, biochemical and physiological) to hypoxia. This study goes further than the 2×2 paradigm mentioned above. It incorporates the use of marker genes both on a population and family basis to search for genetic associations within the wide range of medical, physiological and developmental features under study. Even more interesting is that some of these segregating marker genes are, in fact, polymorphic for enzymes involved as intermediates in the oxygen transport system or tissue utilization of oxygen. The detailed attention given to design, technology, handling of subjects and the logistics of field investigation makes the paper of Schull and Rothhammer of wide utility.

At the Society's symposium, Roger J. Williams recalled his pioneering work in this field as recorded in his classic book *Biochemical Individuality* (John Wiley). It seems fair to conclude from the proceedings of the symposium that while our understanding of physiological variability is still at an early stage, the participants in this symposium have made a number of significant general contributions. They have explored in considerable depth the range of techniques available, though twin, family and epidemiological approaches, and combinations of these. Secondly, the many examples cited show that, for a substantial number of physiological characters, what has been called the "fog of quantitative variation" has begun to disperse.

April 1977   J. S. WEINER

THE GENETIC ANALYSIS OF CONTINUOUS VARIABLES IN MAN

JOHN B. GIBSON

Department of Population Biology, Research School of Biological Sciences,
Australian National University, Canberra

AT the Third International Congress of Human Genetics, Thoday lamented the fact that "we are making very little real progress in the understanding of continuous variation in man" (Thoday, 1967). The remarkable advances that have been made over the last twenty years, particularly in human biochemical genetics and also in many other aspects of the genetics of man, have not been paralleled in the genetic analysis of quantitative traits. Indeed, researchers in other subjects (and here I am thinking particularly of sociologists and psychologists) seeking guidance on the genetics of continuous variables, can be excused for believing that progress since Galton's classic studies (Galton, 1889) provided the basis for human genetics, has been merely honing and polishing a very imperfect model rather than innovatory.

To the extent that this unsatisfactory position derives from the state of the art in organisms experimentally more amenable than man, there is cause for optimism, as recently developed techniques have produced exciting results in *Drosophila*, mice and wheat. There have now been attempts to modify these techniques for use in human populations following the initial stimulus provided by Thoday's paper.

Before briefly outlining the experimental basis of these new techniques, and the modifications to them suggested for human material, it is worth considering some of the reasons why progress has been so slow compared to that in other areas of human genetics.

1

Where phenotypic variation in characters does not fall unambiguously into two or more classes the characters are quantitative rather than qualitative and do not exhibit clear-cut Mendelian segregations. Such normal distributions of phenotypic variation can readily be obtained from the segregation of a large number of genes of similar effects. Indeed, this interpretation formed the basis of multifactorial inheritance and was a necessary assumption for the development of biometrical genetics. Supporters of the concept that a large number of segregating genes must be involved in quantitative inheritance were bolstered by Mather's definition of polygenic characters as those "dependent on the joint action of many genes, each having an effect small in relation to the total non-heritable fluctuation of the character in question" (Mather, 1943). Mather's models stimulated research on quantitative inheritance for they were predictive in relating the genetic architecture to the genetic system. It was demonstrated that polygenes possessed the properties of nuclear-borne genes showing segregation and linkage, but the absence of clear-cut segregations made further genetic analysis difficult. Many assumed from the definition of polygenic characters that they were incapable of analysis on traditional lines and that any analysis lay solely within the area of biometrical genetics, which had developed specifically for the analysis of continuous variables.

However, a continuous distribution does not necessarily indicate the segregation of a large number of genes of small effect. Distributions which do not significantly depart from normality can be obtained from the phenotypic effects of segregation at a single genetic locus (Smith, 1971; Edwards, 1977). Thus the form of the distribution of variation in a character in a population is of little use in determining the number of genes involved, although if pedigree data are available it may be possible to test the goodness of fit of specific genetic models, as suggested by Stewart and Elston (1973) and Elston and Stewart (1973).

In human studies the analysis of continuous variables developed from the theoretical base provided by Fisher (1918), who showed that it was possible to derive measures of resemblance between relatives for polygenic characters as the proportions of genes in common were predictable for various degrees of relationship. The biometrical approach aims to partition the variation first into the genetic and environmental components and then to subdivide the genetic component into that resulting from differences between homozygotes and that resulting from specific effects of various alleles in heterozygotes (Mather and Jinks, 1971).

The interpretation of such analyses applied to human material is fraught with difficulties, for cultural and biological inheritance cannot always be separated and may have similar effects (Feldman and Cavalli-Sforza, 1975). Children in a sibship might resemble their parents because they share similar environments or because they have most of their genes in common with their parents. Resolution of these factors, which may or may not be operating in the same direction for any particular character, is hardly ever possible and familial predisposition by itself is open to a variety of interpretations (Edwards, 1969).

Theoretically the genetic and environmental components of the phenotypic variance can be partitioned by the use of twin studies, but such studies have only rarely been carried out with the most appropriate design (Eaves, 1970, Jinks and Fulker, 1970). Qualitatively, such studies require comparison of sets of monozygotic twins reared together with sets reared apart, as well as control studies on age- and sex-matched sets of dizygotic twins. Quantitatively, Jinks and Fulker (1970) have shown that, even when the genetic component of the variance appears to be as high as 0·6, very large sample sizes are required to convincingly test for significance.

In experimentally amenable organisms some of the difficulties in the biometrical approach can be overcome as environments and matings can be controlled, enabling the biometrical analysis to be made under conditions closely approaching the theoretical ideal. Nevertheless, such analyses cannot lead to the identification of specific genes and hence do not provide material for detailed physiological and biochemical studies with identified genotypes.

Parallel with the development of the biometrical analyses the sophisticated technology of *Drosophila* had been used to detect regions of chromosomes which had effects on specific quantitative characters (Sismanidis, 1942; Wigan, 1949; Breese and Mather, 1957; Hirsch, 1962). Thoday (1961) has extended this technology to provide estimates of the number of effective factors within a region of a chromosome and to locate polygenes with respect to major genetic markers. Essentially the technique for locating polygenes makes use of genetically marked chromosomes which are readily available for *Drosophila melanogaster*. The aim is to determine how many classes of chromosome with respect to their effects on a particular character can be identified after recombination between a marker chromosome and a chromosome affecting the character. In its simplest form, the analysis requires back-crossing an individual heterozygous for a marker chromosome, $m_1 m_2$, and a chromosome affecting the character, $+_1 +_2$, to a marker homozygote, $m_1 m_2/m_1 m_2$ (where m_1 and m_2 represent linked major gene markers and $+_1$ and $+_2$ their wild type alleles, see Fig. 1). The recombinant progeny, $m_1 +_2/m_1 m_2$ and $+_1 m_2/m_1 m_2$, are then assayed for the character and their means compared to those of the parental classes.

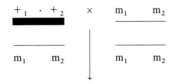

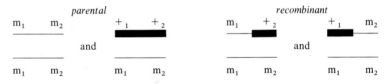

MARKER CLASSES

FIG. 1.   Mating scheme utilizing marker chromosomes to detect region affecting quantitative trait.

Random samples of the recombinant chromosomes are progeny tested by again crossing heterozygotes for a recombinant chromosome to the marker homozygote. Comparison of the results of the progeny tests for the recombinant and parental chromosomes determines how many phenotypic classes, with respect to the parental classes, can be produced by recombination between the two markers (Table 1). Thus if the tester chromosome contains one effective factor in the region between the two markers the recombinant chromosomes will fall into two classes, one of which contains the factor and one which does not. Similarly if there are two effective factors of equal effect in the marked region the recombinant chromosomes will be of three kinds, one similar in mean to the tester chromosome, one similar in mean to the marker chromosome, and the third intermediate in mean between the two parental classes (see Table 1). This basic breeding programme can be modified to remove the markers in the assayed progenies if the markers themselves affect the character being studied.

TABLE 1. Recombinant chromosomes produced, (all heterozygous for marker chromosome).

F = factor affecting trait.
f = constitution of marker chromosome at trait locus.

ONE factor		
$+_1$ F $+_2$	m_1 F $+_2$	m_1 f $+_2$
m_1 f m_2	$+_1$ F m_2	$+_1$ f m_2

TWO factors			
$+_1$ F_1 F_2 $+_2$	m_1 F_1 F_2 $+_2$	m_1 f_1 F_2 $+_2$	m_1 f_1 f_1 $+_2$
m_1 f_1 f_2 m_2	$+_1$ F_1 F_2 m_2	$+_1$ F_1 f_2 m_2	$+_1$ f_1 f_2 m_2

trait phenotypic class	similar to heterozygous parent	Intermediate	similar to marker homozygous parent

These techniques, and others based on the same principles, have achieved remarkable successes in *Drosophila*. In one analysis the located polygenes accounted for over 80 per cent of the differences accumulated in a directional selection line (Spickett and Thoday, 1966) and provided detailed information on the developmental basis of a quantitative character (Spickett, 1963).

In mice (Shire, 1969) and wheat (Wehrhahn and Allard, 1965; Law, 1966) a similar use of marker chromosomes has been successfully exploited to dissect the genetic components of a variety of continuous variables and has shown that effective factors can be identified, located and account for a substantial proportion of the genetic variance.

These studies have stimulated discussion on the number and

nature of genes affecting continuous variables. Arguments concerning the number of genes involved are not very productive for this must depend on the developmental complexity of the character. Phenotypes studied are often far removed from initial gene action and conceal a variety of structural gene systems modified by controlling processes. Leaving aside this aspect of the discussion the important point to emphasize is that these studies have identified factors which contribute significantly to the genetic variance in continuous variables and can be located in the genome by using segregating marker genes.

There is clearly a vast gulf between this kind of work in experimentally amenable organisms and man. However, once it is recognized that a proportion, sometimes a substantial proportion, of the genetic variance in continuous variables in human populations may be due to segregation at a few loci with relatively large effects, an experimental approach can be formulated. The location of polygenes technique makes use of marker genes which are now available in human populations primarily in the form of polymorphisms with alleles at intermediate frequencies, although the techniques must be modified for human material where neither environments nor matings can be controlled.

Penrose (1934 and 1938) first considered the problem of detecting linkage between a quantitative trait and a marker locus from sib pair data when having no knowledge of their parents. He failed to find significant linkage between eye colour and ABO blood groups in fifty pairs of sibs and suggested that samples in excess of 100 pairs would be required, although this number now appears to be far too few.

Thoday (1967), Jayakar (1970) and Haseman and Elston (1972) have all suggested methods by which studies of segregating marker genes can be used to investigate linkage between continuous variables and marker loci. Thoday (1967) outlined an analysis combining segregation of marker genes and measurements on continuous variables in families in which both parents were heterozygous for a marker locus. He used as an example segregation at the MN locus and considered a model in which a two allele locus affecting a continuous variable was completely linked to the MN locus and in linkage phase equilibrium in the population. The continuous variable in such a case would have a higher variance within and a low variance between families in the MN than the M or N individuals of the progenies of MN × MN matings. Sib pair analysis should reveal lower within sib pair correlations for MN pairs than for M pairs or N pairs.

The methods suggested by Jayakar (1970) for the detection of linkage by using the ratio of variances require at least two members of one marker genotype and one of another within each family. Thus the analysis requires data on large families which is a serious restriction for human studies.

Haseman and Elston (1972) described procedures using data on sib pairs for estimating linkage between a known marker locus and a two allele locus affecting a quantitative trait. The technique makes use of all types of mating with respect to the marker locus and requires data for the marker locus on both parents and two sibs. As secular and age effects could introduce large biases into measurements of quantitative traits taken on two or more generations the analysis is restricted to data for sib pairs. Haseman and Elston (1972) provide a non-parametric method of detecting linkage in such data and outline a maximum likelihood method of estimating linkage. The technique has been tried on a sample of 89 dizygotic same sex Norwegian twin pairs (Elston, Kringlen and Namboodiri, 1973). It revealed a significant relationship between the *Gc* locus and "psychosis in general" (as revealed by questionnaire), although the authors urge that the results be interpreted with caution.

Inevitably there is uncertainty regarding the sample sizes required for such studies if they are to reveal linkages. However it is clear that the required sample sizes will vary with gene frequency and with the proportion of variance accounted for by the linked polygenic locus (loci). Both Renwick (1973) and Robertson (1973) have argued that the chance of detecting a linkage using this method is extremely low, although Elston (1973) has countered that this view is unduly pessimistic. The more markers and traits that are available for the sample the larger the prior probability of there being a linkage within detectable mapping distance. Elston (1973) has shown that if you have two marker loci and ten trait loci (or the converse) randomly distributed throughout the human autosomes, the probability that at least one marker and one trait loci should be within 44 centimorgans is about 0·39. The probability of detecting this linkage is only reasonable using sib pairs if the actual distance between the two loci is very small. However, Blackwelder and Elston (1974) have shown that if the genetic variance contributed by the locus is equivalent to a heritability of 50 per cent then the necessary sample size is 265 sib pairs.

Nevertheless, there are other reasons for cautious optimism that it will be possible to identify genes of some continuous variables. Bock and Kolakowski (1973) have argued that segregation at an X-linked locus accounts for about 45 per cent of the variability in human spatial ability. Their arguments are based on data which show that for this trait the father–son correlation is near zero, the father–daughter correlation is similar to that of the mother–son, and the mother–son correlation is greater than that for mother–daughter. (See also Guttman, 1974 and Yen, 1975). This pattern is consistent with that expected for X-linkage with complete inheritance (Table 2). It is also relevant that spatial ability is related to components of throwing accuracy (Kolakowski and Malina, 1974)

TABLE 2. Comparison of familial correlations for spatial ability, stature and sex linked IgM concentration. (Data taken from Bock and Kolakowski, 1973.)

Trait	Familial correlations			
	Father–son	Father–daughter	Mother–son	Mother–daughter
Spatial ability	0·11	0·28	0·25	0·14
IgM concentration	0·10	0·31	0·36	0·19
Stature	0·51	0·51	0·49	0·51

and perhaps might be expected to have evolved a sex linked mode of inheritance. Tests of hypotheses concerned with sex linkage are likely to require smaller sample sizes than those concerned with autosomal inheritance. In the Otmoor study of Harrison *et al.* (1974), data were available on the spatial component of I.Q. and the Xg blood group status of 32 male sib pairs. These data have been analysed using a minor modification of the sib pair method suggested by Penrose. The sib pairs were divided into those concordant and those discordant for Xg status and then each of these categories were further divided on the basis of whether both members of the pair were similar for spatial ability scores. This was done simply by dividing the distribution of scores into two equal halves and labelling one "high" and the other "low". The resulting χ_1^2 of 4·4 (0·05 > P > 0·01) suggests linkage between the Xg locus and factors affecting spatial ability. When further sib pair samples, that are now available from the same population, have been analysed it should prove possible to estimate the linkage with respect to the Xg locus.

The analysis outlined above was substantially aided because at least a component of the trait was sex-linked but spatial ability is a complex trait and is likely to have a number of phenotypic components. Spickett, Shire and Stewart (1967) demonstrated very nicely the increased precision in genetic analysis that can be obtained by dissecting a character into its component parts. They showed that the distinction between two mouse strains in corticosteroid hormone per unit of body weight increased when they focused attention on the quantity of a particular hormone per unit weight of hormone-producing tissue. In fact, after redefining the character they were able to explain the difference between the two strains in terms of a one locus hypothesis as the F_2's segregated into three discontinuous groups.

Results such as these strongly suggest that a profitable approach to the genetic analysis of continuous variables in human populations would be to exploit character analysis. Indeed attention needs to be focused on variation in the enzymes that one suspects may be involved in any particular trait for all genes must be responsible for discontinuous variation if we can get close to the initial gene products.

Two examples will serve to illustrate this approach. The first shows very neatly that a normal phenotypic distribution does not necessarily imply a large number of genes. Spencer, Hopkinson and Harris (1964) and Eze, Tweedie, Bullen, Wren and Evans (1974) have shown that about 60 per cent of the variation in red blood acid phosphatase activity in the population could be accounted for by the segregation of three alleles at a single structural gene locus. The activity of the enzyme is normally distributed because of the frequencies of the three alleles in the population and because there is variation in activity within each of the six genotypes.

The second example, concerned with susceptibility to lung cancer, demonstrates both the value of character redefinition and the rewards of concentrating attention on a readily assayable enzyme-system. The enzyme aryl hydrocarbon hydroxylase is a mixed function oxidase in the microsomal fraction of many mammalian tissues. Although its precise physiological function is unknown, it is known to hydroxylate a variety of exogenous hydrocarbons, including those in tobacco smoke, forming carcinogens. The enzyme can be assayed in cultured lymphocytes (Busbee, Shaw and Cantrell, 1972), where the level of enzyme activity forms a normal distribution in the population. However, the enzyme is inducible by the hydro-carbon substrates and the induced levels in the population form a trimodal distribution. Family studies have indicated that the distri-bution of the induced levels is compatible with the hypothesis of segregation of two alleles at a single genetic locus (Kellermann, Luyten-Kellermann and Shaw, 1973). The postulated three geno-types AA, AB and BB represent the low, medium and high inducible phenotypes respectively. It has been shown that the frequency of the high inducible genotype is greater in a sample of patients diag-nosed lung cancer than in the population at large (Kellermann, Shaw and Luyten-Kellermann, 1973).

These observations are potentially extremely important and suggest that other complex characters might be amenable to analysis along similar lines. Recent findings on the mechanism of the inhi-bition of 3-hydroxy-3-methyl-glutaryl coenzyme A reductase, the rate-controlling enzyme in the cholesterol biosynthetic pathway, (Brown and Goldstein, 1974) indicate that variation in serum cholesterol levels might be worth analysing in material where information is available on segregating markers. It has been sug-gested (Mayo, Wiesenfield, Stamatoyannopoulos and Fraser, 1971) that the association between serum cholesterol levels and the ABO locus may be due to pleiotropy as the association has been observed in many geographically separate populations. Further, as the ABO locus contributes to the variation in serum trehalase levels, it would be informative to follow segregation at these marker loci in family studies on serum cholesterol levels.

It should soon be possible to critically test some of the new tech-niques for analysing quantitative traits as the required data sets are being collected in a number of studies on human populations. Preliminary reports have been published on some of the studies and further results are awaited with interest (De Fries, Vandenberg,

McClearn, Kuse, Wilson, Ashton and Johnson, 1974).
Let us hope that such work combining studies of quantitative traits and segregating marker loci will be as rewarding in human material as it has been in lower organisms and that like Robertson (1967) we will find that "the fog of quantitative variation, which I had assumed to be the result of segregation at a large number of loci, has gradually begun to clear as I have recognized individual segregations almost as personal friends".

References

BLACKWELDER, W. C. and ELSTON, R. C. (1974) Comment on Dr Robertson's Communication. *Behav. Genet.*, **4**, 97–99.

BOCK, R. D. and KOLAKOWSKI, D. (1974) Further evidence of sex-linked major-gene influence on human spatial visualizing ability. *Am. J. Hum. Genet.*, **25**, 1–14.

BREESE, E. L. and MATHER, K. (1957) The organization of polygenic activity within a chromosome in *Drosophila*. I. Hair characters. *Heredity*, **11**, 373–395.

BROWN, M. S. and GOLDSTEIN, J. L. (1974) Expression of the familial hypercholesterolemia gene in heterozygotes: Mechanism for a dominant disorder in man. *Science*, **185**, 61–63.

BUSBEE, D. L., SHAW, C. R. and CANTRELL, E. T. (1972) Aryl hydrocarbon hydroxylase induction in human leucocytes. *Science*, **178**, 315–316.

DE FRIES, J. C., VANDENBERG, S. G., McCLEARN, G. E., KUSE, A. R., WILSON, J. R., ASHTON, G. C. and JOHNSON, R. C. (1974) Near identity of cognitive structure in two ethnic groups. *Science*, **183**, 338–339.

EAVES, L. J. (1970) The genetic analysis of continuous variation: a comparison of designs applicable to human data. II. Estimation of heritability and comparison of environmental components. *Br. J. Math. Statist. Psychol.*, **23**, 189–198.

EDWARDS, J. H. (1969) Familial predisposition in man. *Br. med. Bull.*, **25**, 58–64.

EDWARDS, J. H. (1977) The analysis of general inheritance. This volume, pp. 15–21.

ELSTON, R. C. (1973) Reply to "Message from a referee on the Elston method". *Behav. Genet.*, **3**, 319–320.

ELSTON, R. C., KRINGLEN, E. and NAMBOODIRI, K. K. (1973) Possible linkage relationships between certain blood groups and schizophrenia or other psychoses. *Behav. Genet.*, **3**, 101–106.

ELSTON, R. C. and STEWART, J. (1973) The analysis of quantitative traits for simple genetic models from parental, F_1 and backcross data. *Genetics*, **73**, 695–711.

EZE, L. C., TWEEDIE, M. C. K., BULLEN, M. F., WREN, P. J. J. and EVANS, D. A. P. (1974) Quantitative genetics of human red cell acid phosphatase. *Ann. Hum. Genet., Lond.*, **37**, 333–340.

FELDMAN, M. W. and CAVALLI-SFORZA, L. L. (1975) Models for cultural inheritance: a general linear model. *Ann. Hum. Biol.*, **2**, 215–226.

FISHER, R. A. (1918) The correlation between relatives on the supposition of Mendelian inheritance. *Trans. Roy. Soc. (Edinburgh)*, **52**, 399–433.

GALTON, F. (1889) *Natural Inheritance*. Macmillan and Co., London.

GUTTMAN, R. (1974) Genetic analysis of analytical spatial ability: Raven's progressive matrices. *Behav. Genet.*, **4**, 273–284.

HARRISON, G. A., GIBSON, J. B., HIORNS, R. W., WIGLEY, J. M., HANCOCK, C., FREEMAN, C. A., KÜCHEMANN, C. F., MACBETH, H. M., SAATCIOGLU, A. and

CARRIVICK, P. J. (1974) Psychometric, personality and anthropometric variation in a group of Oxfordshire villages. *Ann. Hum. Biol.*, **1**, 365–381.

HASEMAN, J. K. and ELSTON, R. C. (1972) The investigation of linkage between a quantitative trait and a marker locus. *Behav. Genet.*, **2**, 3–19.

HIRSCH, J. (1962) Individual differences in behaviour and their genetic basis. In *Roots of Behaviour*, Ed. E. L. Bliss, pp. 3–23. P. Hoeber, New York.

JAYAKAR, S. D. (1970) On the detection and estimation of linkage between a locus influencing a quantitative character and a marker locus. *Biometrics*, **26**, 451–464.

JINKS, J. L. and FULKER, D. W. (1970) Comparison of the biometrical genetical, MAVA, and classical approaches to the analysis of human behaviour. *Psychol. Bull.*, **73**, 311–349.

KELLERMANN, G., LUYTEN-KELLERMANN, M. and SHAW, C. R. (1973) Genetic variation of aryl hydrocarbon hydroxylase in human lymphocytes. *Am. J. Hum. Genet.*, **25**, 327–331.

KELLERMANN, G., SHAW, C. R. and LUYTEN-KELLERMANN, M. (1973) Aryl hydrocarbon hydroxylase inducibility and bronchogenic carcinoma. *New Engl. J. Med.*, **289**, 934–937.

KOLAKOWSKI, D. and MALINA, R. M. (1974) Spatial ability, throwing accuracy and man's hunting heritage. *Nature, Lond.*, **251**, 410–412.

LAW, C. N. (1966) The location of genetic factors affecting a quantitative character in wheat. *Genetics*, **53**, 487–498.

MATHER, K. (1943) Polygenic inheritance and natural selection. *Biol. Rev.*, **18**, 32–64.

MATHER, K. and JINKS, J. L. (1971) *Biometrical Genetics: The Study of Continuous Variation*. 2nd edition. Chapman and Hall, London.

MAYO, O., WIESENFELD, S. L., STAMATOYANNOPOULOS, G., and FRASER, G. R. (1971) Genetical influences on serum cholesterol level. *Lancet*, **ii**, 554–555.

PENROSE, L. S. (1934) The detection of autosomal linkage in data which consist of pairs of brothers and sisters of unspecified parentage. *Ann. Eugen.*, **6**, 133–138.

PENROSE, L. S. (1938) Genetic linkage in graded human characters. *Ann. Eugen.*, **8**, 233–338.

RENWICK, J. H. (1973 Message from a referee on the Elston Method. *Behav. Genet.*, **3**, 317–318.

ROBERTSON, A. (1967) The nature of quantitative genetic variation. In *Heritage from Mendel*, Ed. R. A. Brink and E. D. Styles, pp. 265–280. The University of Wisconsin Press, Madison, Wisconsin.

ROBERTSON, A. (1973) Linkage between marker loci and those affecting a quantitative trait. *Behav. Genet.*, **3**, 389–391.

SHIRE, J. G. M. (1969) Genetics and the study of adrenal and renal function in mice. In *Progress in Endocrinology*. Ed. C. Gual, pp. 292–296. (Proceedings of the Third International Congress of Endocrinology.) Excerpta Medica, New York/Amsterdam.

SISMANIDIS, A. (1942) Selection for an almost invariable character in *Drosophila*. *J. Genet.*, **44**, 204–215.

SMITH, C. (1971) Discriminating between different modes of inheritance in genetic disease. *Clin. Genet.*, **2**, 303–314.

SPENCER, N., HOPKINSON, D. A. and HARRIS, H. (1964) Quantitative differences and gene dosage in the human red cell acid phosphatase polymorphism. *Nature, Lond.*, **201**, 299–300.

SPICKETT, S. G. (1963) Genetic and developmental studies of a quantitative character. *Nature, Lond.*, **199**, 870–873.

SPICKETT, S. G., SHIRE, J. G. and STEWART, J. (1967) Genetic variation in adrenal and renal structure and function. *Mem. Soc. Endocrin.*, **15**, 271–288.

SPICKETT, S. G. and THODAY, J. M. (1966) Regular responses to selection. III Interaction between located polygenes. *Genet. Res.*, **7**, 96–121.

STEWART, J. and ELSTON, R. C. (1973) Biometrical genetics with one or two loci: the inheritance of physiological characters in mice. *Genetics*, **73**, 675–693.

THODAY, J. M. (1961) Location of polygenes. *Nature, Lond.*, **191**, 368–370.

THODAY, J. M. (1967) New insights into continuous variation. In *Proc. Third Int. Cong. Hum. Genet.*, Ed. J. F. Crow and J. V. Neel, pp. 339–350. John Hopkins, Baltimore.

WEHRHAHN, C. and ALLARD, R. W. (1965) The detection and measurement of the effects of individual genes involved in the inheritance of a quantitative character in wheat. *Genetics*, **51**, 109–119.

WIGAN, L. G. (1949) The distribution of polygenic activity on the X chromosome of *Drosophila melanogaster*. *Heredity*, **3**, 53–66.

YEN, W. M. (1975) Sex-linked major-gene influences on selected types of spatial performance. *Behav. Genet.*, **5**, 281–298.

THE ANALYSIS OF GENERAL INHERITANCE

J. H. EDWARDS

The Infant Development Unit, Queen Elizabeth Medical Centre, Birmingham

THE condition of an organism may be expressed in terms of states and variates: a male diabetic with a defined height, weight and resting glucose level, can be defined as

$$M, d; 170 \cdot 3, 64 \cdot 1, 190 \cdot 7.$$

These are only a few of the sets of states and variates from which any evaluation, diagnosis or prognosis is made, and the essence of both physiological and pathological enquiry is to direct attention, in the first instance, to those states and variates most relevant to the matter at issue. States are of two general forms, those which merely summarize a range of some variate, as in such adjectives as tall, intelligent or hypertensive, and those which specify a primary attribute which differ qualitatively or discontinuously from other states: for example, male specifies a state which in normal usage is discontinuous from its obverse, non-male or female. The distinction is not always possible, in that some attributes, such as having red hair, are not discontinuous, although intermediates may be relatively uncommon.

States may be related to the constitution, or genetic make-up, in three ways. They may be simply defined by some threshold in a continuous distribution, as in "tall", "short", "fat", or "intelligent", and such distributions are obviously to some extent influenced by inborn factors. Secondly, states may be the consequence of some underlying continuous distribution which leads to discontinuous

consequences. For example, variations in bone strength will predispose some to fracture and variations in arterial pressure will
predispose to arterial rupture. Variations in developmental punc-
tuality may lead to a total failure of various structures to form
properly. Variations in the body's defences may lead to infective
disease in those most poorly defended.

Finally, the discontinuity may merely reflect a simple genetic
discontinuity, as in such Mendelian disorders as achondroplasia,
albinism, or haemophilia.

The former two cases are sometimes referred to as multifactorial
or polygenic. In the former word various authors exclude or include
environmental factors; polygenic is sometimes used synonymously,
sometimes with the added implication that all the genetic factors
arc individually small, and sometimes with the implicit exclusion of
polyallelism. In fact, the exclusion of even one strong allele, that is,
of an allele substantially influencing the phenotype, cannot be
inferred in man from the scale and quality of data normally available.

The Variate Phenotype

A phenotype expressed entirely in variate terms, without reference
to states, is easier to handle. Each individual can be expressed as a
point in a coordinate system, and each family as a set of points, or,
more generally, a set of points in a space of $n \times m$ dimensions
where there are n variates and m family members.

In the simplest case we can consider the heights of fathers and
sons studied by Galton, Pearson and others. In a three-dimensional
space we could represent fathers, mothers and sons.

Such a space has a number of simple features, and, provided the
distributions and their linear functions are normally distributed, it
can be summarized quite simply by the sums of the points measured
from each plane, and the sums of the products of each pair of
measurements, including the measurement with itself. That is, the
sums

$$Sx_i, \ Sy_i, \ Sz_i \ \text{and} \ Sx_ix_i, \ Sy_iy_i, \ Sz_iz_i, \ Sx_iy_i, \ Sx_iz_i, \ Sy_iz_i$$

The simplest biometric argument is to regard the genotype as
blurred in its expression, in the same way as a photographic image

may be blurred by poor focusing and to relate this degree of blurring to the environment on the ground that what is not genetic is environmental. However, this definition by default is not the environment of the environmentalist, which is what can be controlled or modified. The matter is of some importance, since it is sometimes assumed that as heritability increases environmental manipulation becomes less effective. In fact, very high heritabilities may be associated with necessary environments, as in exposure to reading, bicycling or tuberculosis being necessary for the manifestation of an inborn aptitude, and such environments can be modified.

Improved nutrition makes people larger, without necessarily influencing the correlation between relatives, if evenly applied. If unevenly applied the familial concentration of privilege and poverty are likely to increase the phenotypic correlation, giving a high phenotypic correlation and, if the ratio of this to the genetic correlation is estimated, a curiously high heritability. For this reason it is wiser to use the term "familiarity" in man, and not to prejudge the issue by the word heritability.

Such measures tell us nothing about the nature of the hereditary determinants; indeed, provided the allelic effects are additive, the phenotypic correlation is formally irrelevant to the numbers and strengths of these determinants. Strong determinants may show up by a discontinuity, so that a single locus may be responsible for great deal of the variation. However, such single loci may be very difficult to detect when the phenotype is loosely clothed in other sources of genetic variation. If we take a simple example, such as height in adults, for practical purposes the Y chromosome acts as though it were an allele which will add about 1·7 units of standard deviation to the structure. However, without supplementary aids to defining sex, it would not be easy to infer the presence, or to estimate the extent, of this discontinuity.

Evidence of such discontinuities may be sought in the segregation of phenotypes, but here, as with height and maleness, unless some other factor can be seen to segregate it will be difficult to find evidence from the distribution of the measurements, and elaborate methods seem unlikely to be much more powerful than unaided observation. This is easily seen from Figs. 1 to 3, which shows the distribution of all possible pairs of sibs for various strengths of allele.

In Fig. 1 the distributions do not overlap: all genotypes are

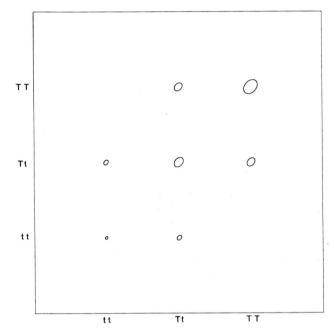

FIG. 1.   Distribution of tallness in pairs of first degree relatives where tallness is almost entirely defined by two alleles of intermediate effect and unequal frequency ($t = 0.2$). The ellipses are contours of the normal surface.

distinct and there would be no problem in defining the determinants. This sort of situation arises with some laboratory observations, for example the MN blood groups, but it is unusual to be able to distinguish all the genotypes in the flesh.

In practice, as the effects of the strongest locus become weaker, we get the situation we find in Figs. 2 and 3, the genotypes becoming buried in the variation imposed by other loci.

As these vary, different situations will give different phenotypic correlations, or heritabilities, and, as stated earlier, such measures cannot help in the identification of strong alleles or influential loci.

These measures allow the correlation coefficients between the phenotypes to be derived. Since it is known that, in a uniform environment in which non-interacting genetic units of additive effect segregate evenly and independantly, the phenotypic consequences of these determinants will have a correlation coefficient of

$(\frac{1}{2})^n$ where n is the degree of relationship (1 for sibs, children, and parents; 2 for nephews, nieces, uncles and aunts, 3 for first cousins) a comparison of the correlation found with that expected may be used to estimate the "strength" or effectiveness of the genetic determinants.

A particularly simple case arises with identical twins, or relatives of zero degree, since any difference will usually be due to a failure of these determinants to express themselves. Non-identical twins are sometimes used as controls, although this is not usually necessary.

The ratio of the correlation observed to that expected is sometimes called the heritability by analogy with animal breeding, in which it is highly correlated with speed of response to selection, and is of practical value in the strategy of a rapid response.

The documentation of the great similarity of identical twins has been of major interest in, firstly, allowing studies of the effects of

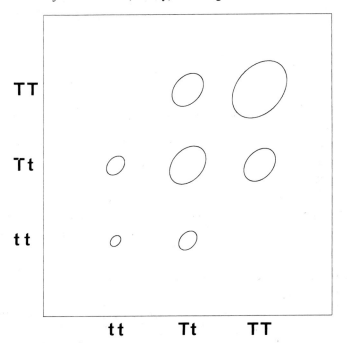

FIG. 2. Distribution of tallness is pairs of first degree relatives where tallness is largely defined by two alleles of intermediate effect and unequal frequency ($t = 0.2$). The ellipses are contours of the normal surface.

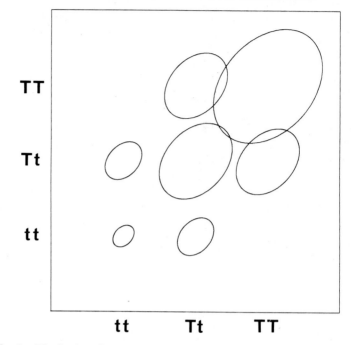

Fig. 3. Distribution of tallness in pairs of first degree relatives where tallness is partly defined by two alleles of intermediate effect and unequal frequency ($t = 0.2$). The ellipses are contours of the normal surface.
Such distributions would be difficult to distinguish from those due to numerous very weak determinants.

distinct environments, as following separate adoption, or differential exposure to intellectual or athletic training, or to malnutrition. Such data are readily portrayed as points, each point representing some value in one twin along one axis and in the other along the other axis. If non-identical twins are similarly displayed, they are immediately seen to be represented by a wider scatter of points about the main diagonal.

Such distributions inform us of the limitations of differing environments, or of the strength of the constitution to manifest themselves, but it is difficult to make any more exacting interpretation of the many lucid and informative demonstrations of twin similarity.

If phenotypic measurements either in populations or families fail to inform us of the presence of major alleles, is there any other approach? The only approach which has been fruitful so far, or, indeed, which would be expected to be fruitful, is the detection and interpretation of associations between markers, by which is usually meant qualitative genetic variants, such as sex or blood group, with some measurement, or some trait, such as being diabetic.

Associations are usually consequent on functional relationships, as those relating sex to height and aggressiveness, which seem to be largely mediated by a single hormone, or the association of gastric disease with the swallowing of blood group substances in saliva, or of skin pigmentation with sunburn. Associations with markers which exist in families, but not in populations, are usually the consequence of their determinants being conveyed in proximity on the same chromosome, or being linked. Very close linkage may lead to associations in populations, as in those found in the HLA loci, or between alleles at one locus and disorders predisposed to by alleles at neighbouring loci.

The last method has been particularly powerful in defining strong alleles predisposing to such disorders as coeliac disease, juvenile diabetes, and ankylosing spondylitis, although these disorders had previously all been explicable as the consequence of large numbers of loci of small influence.

Unfortunately, the population method will only reveal very close linkage, and family methods require very extensive data, usually involving hundreds of individuals, if any sound conclusions are to be drawn, as well as a homogeneity in the genetic basis of the trait or measurement under discussion.

In summary, numerical methods of data collection lend themselves naturally to both graphical display and numerical analysis. The latter, being restricted to the fundamental operations of addition, subtraction, multiplication and division, is only powerful when the measures being operated upon are the consequence of essentially identical determinants, as in models which assume that polygenes are entities. They have limited analytic power for the detection of the distinct determinants, and the recognition or uncovering of such determinants would seem to be the most fertile approach in the analysis of numerical, as well as of qualitative, data. At present the use of complex numerical summaries of data has little to commend it over the use of graphical display of data, with or without simple methods of averaging, such as regression lines or incidences, and simple measures of precision, such as standard deviations.

In anthropology there are serious difficulties in collecting data from large families and, where extensive data have been collected, as in the same major studies on populations in South America, even the most detailed and exacting methods of analysis have yielded only a disappointingly small addition to the extensive knowledge gained by simpler approaches.

The various approaches available for such studies have recently been brought together and discussed against a background of genuine data (Cavalli-Sforza and Bodmer, 1971). No doubt further methods of analysis will be discovered, but it seems doubtful if they will revolutionize what can be done with such data, although improvements in storage, recall, and display will greatly assist informed interpretation (Edwards, 1969).

References

CAVALLI-SFORZA, L. L. and BODMER, W. F. (1971) *The Genetics of Human Populations*. W. H. Freeman & Co. San Francisco.

EDWARDS, J. H. (1969) Familial predisposition in man. Reprinted from *British Medical Bulletin*, Vol. 25, No. 1 (New Aspects of Human Genetics), pp. 58 64.

METHODS AND PROBLEMS IN
PHYSIOLOGICAL GENETICS

D. F. ROBERTS

Department of Human Genetics, University of Newcastle upon Tyne

THE experimental requirements for seeking and quantifying genetic components in physiological variation may be complex, and other contributors have shown the difficulty of identifying such components in almost all the variables so far discussed. This is partly due to the nature of the variables themselves, characterized by lability and relative rapidity of response, but also may be partly due to the extent to which these investigations have relied on twin analysis, which taken alone poses considerable interpretative difficulty. But population genetics has devised other methods by which such estimates can be made, and these and the problems that arise in them can be illustrated by some of our own studies in the Newcastle Human Genetics Department on physiological variables. But first some general theoretical comments on the genetic basis of physiological variation.

Genetic Basis

There is relatively little direct evidence on the genetic component in normal physiological variation. A few single genes are known to produce a pronounced effect, for example on the threshold of tasting sensitivity to PTC. There are instances of an all or nothing response (an ability to respond perhaps), as in the inability to sweat that comes in anhydrotic ectodermal dysplasia, which are under or remain candidates for monogenic control, while departure of some measure outside the normal physiological range is often the first detectable sign of the presence of a disease-producing gene or genes, e.g. elevated blood pressure in the adult form of polycystic kidney, the

23

glucose-tolerance test in diabetes. Apart from such clinical conditions, any genetic component to normal physiological variation that may exist seems to be most reasonably understood on a polygenic hypothesis. Polygenes are transmitted in the same way as, and in accordance with the same laws as, major genes, but their effects do not provide sufficient discontinuity for individual study. A polygene acts as one of a system, the members of which may act together or against each other respectively to effect large phenotypic differences or inhibit them. An individual polygene has only a slight effect; it is apparently interchangeable with others within the system; it does not have an unconditional advantage over its allele, since its advantage depends on other alleles present in the system; and it cannot therefore be heavily selected for or against, that is not until its associates in the same system come to be linked with it. The stability that this gives is of obvious evolutionary importance, for it means that evolution acts not on the ability to respond (without which disaster may occur) but on the level of response (survival but at different levels of efficiency), and so it would not be surprising to find that this mode of genetic control applies to fundamental features critical for survival and individual efficiency, such as physiological responses to varying external and internal conditions.

This does not mean loss of plasticity, for in polygenic inheritance similar phenotypes develop from different genotypes, and therefore great genetic diversity and potentiality for change is concealed behind the phenotypic variation. The result in the individual depends more on the numbers of genes making for increased measurement of the variable than on the particular genes present. Fig. 1 shows the variation that would be produced in a quantitative character such as normal blood pressure if three pairs of allelic genes were responsible for it, the gene for increased blood pressure being denoted by the capital letter in each case. It is assumed that the genes are all equal in their effect, that they simply add to each other, and that the alleles of each pair are equally frequent. With more genes, the irregularities of the steps of the histogram disappear, and the curve becomes smooth. Thus a normal curve of distribution of a quantitative character in a population may be attained entirely by genetic determination. If such a population were uniformly exposed to a different environment increasing the character, the whole curve would shift to the right. If only part of the population were so

THREE LOCI, TWO ALLELES AT EACH, EQUAL AND ADDITIVE IN EFFECT: DISTRIBUTION OF GENOTYPES

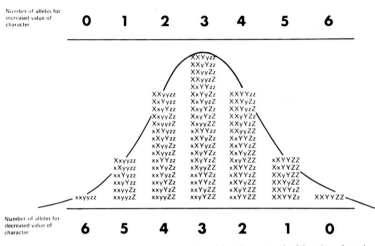

Number of alleles for increased value of character:	**0**	**1**	**2**	**3**	**4**	**5**	**6**

FIG. 1.   Variation in a quantitative character under the control of 3 pairs of equifrequent alleles at unlinked loci.

exposed, the variability of the curve would increase (Fig. 2).

The extent of the genetic contribution to such quantitatively varying characters in a population can be measured by partitioning the variance. In essence, total phenotypic variance is divided into that due to (1) additive polygenic effects, (2) environmental effects, and (3) other factors. The additive genetic contribution expressed as a proportion of the total variance is known as the heritability. A character that is totally genetically determined would have a heritability of 100 per cent; that for which all the variation is environmental would have a heritability of zero. Heritability can be measured in a number of ways, for example from the examination of the resemblance between relatives of different degrees who therefore have different proportions of their genes in common. First degree relatives have 50 per cent of their genes in common, second degree 25 per cent, third degree $12\frac{1}{2}$ per cent; these figures indicate the degree of resemblance to be expected were all the variation of genetic origin, so that observed resemblances can be compared with these to indicate the relative contributions of genetic and environmental components. A different estimate, also confusingly known as heritability, comes from twin studies alone. But even if for a given feature

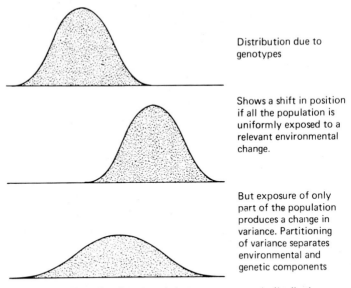

Distribution due to genotypes

Shows a shift in position if all the population is uniformly exposed to a relevant environmental change.

But exposure of only part of the population produces a change in variance. Partitioning of variance separates environmental and genetic components

FIG. 2.   Effect of environmental change on a genotypic distribution.

one establishes the genetic contribution to the variation within a population as being, say, 100 per cent, this implies nothing as to the genetic basis of differences between populations. Theoretically it is possible for variation within a population to be totally genetic and for variation between populations to be purely environmental.

At the moment therefore, it is not possible to quantify the genetic contribution to any physiological difference between populations, for the correct experiments, intricate as they are and involving hybrids as well as random families, have not been carried out. Many of the physiological differences that at one time were regarded as racial have been shown to be due to variations in technique of measurement or to the differing environments in which the groups live (e.g. Roberts, 1952); for very few, after these effects are taken into account, does there remain any suggestion of population differences that may be genetic in origin. This is not to say that they do not exist for, after all, considerable differences in gene frequencies are well established for human populations for monogenic characters such as blood groups and isoenzyme types (Mourant et al., 1976), there is a well established genetic basis to differences in skin colour (Harrison and Owen, 1964; Kahlon, 1973), and the probability is high that the striking differences in some body proportions owe much to genetic differences. Certainly physiological variation is an almost untouched field of genetic investigation that deserves at least exploratory cultivation, and perhaps this will develop after intrapopulation enquiry has established which physiological variables are likely to yield a profitable harvest.

Simple Correlation—Thyroid Function

Simple correlations between relatives provide useful pointers to the characters in which investigation is worth pursuing further, and this can be illustrated by reference to thyroid function. Among its several functions, the thyroid gland controls the rate of metabolism, oxygen consumption and thereby heat production, and also influences normal skeletal growth and development. This control is exercised by the secretion into the blood of the hormones thryroxine (T_4) and triiodothyronine (T_3), the latter also being produced by peripheral conversion from T_4. In the thyroid, these hormones are stored bound to a glycoprotein, thyroglobulin. The iodine required for their synthesis is obtained from the diet (ultimately from the soil from which the foodstuffs originate) and circulates in very small amounts as inorganic iodide ($< 1 \mu g/100$ ml). Most of the iodine in the blood is present as T_4 and T_3 bound to carrier proteins and can be measured as protein-bound iodine (PBI, normal range 3–8 $\mu g/100$ ml). The thyrotrophic hormone (TSH) from the pituitary is responsible for the maintenance of normal thyroid function, its secretion being controlled by the level of circulating thyroid hormone. With the recent development of methods of immunoassay, the normal circulating levels of TSH, T_4 and T_3 can be measured.

These and other measures of thyroid function have been included in several of our recent family studies. So far the data are sufficient only for a preliminary examination of similarities between first-degree relatives, and are scanty for comparisons of relatives of more remote degree. All the figures relate to apparently normal individuals, with no sign of thyroid disease, so that none of the single major genes that can cause interference in the thyroid pathways are present. Table 1 shows the correlation coefficients between pairs of first-degree relatives from 32 different families. Significant and moderately high correlation coefficients between first-degree relatives occur for PBI and T_3. The lower figure for T_4, using the same pairs of relatives as the T_3 estimates, may indicate greater technical variation in measurement, it may merely be a chance effect due to the small numbers examined, or it may be real; if real, in view of the close biochemical relationship between the two substances (the molecules differing in only a single iodine atom) it is unlikely to represent a difference in genetic control of mechanisms of synthesis, but a possible explanation may lie in the increased rate of peripheral disposal of T_4, for instance in exposure to cold. Certainly these three results suggest that further investigation may prove rewarding. By contrast, the lower figures for TSH and TSI suggest that these substances may be of less interest for genetic analysis. The finding for the thyroid stimulating immunoglobulin level is not surprising in the light of what follows (pp. 10–12) but there remains the problem of why the TSH correlation appears so different from the other hormone levels.

TABLE 1.   Intrafamilial correlations of thyroid function measures.

	1st degree relatives	
	n (pairs)	r
Protein-bound iodine (PBI)	40	+ 0·498**
Thyroid stimulatory hormone (TSH)	20	+ 0·154
Triiodothyronine (T_3)	33	+ 0·478**
Thyroxine (T_4)	33	+ 0·258
Thyroid stimulatory immunoglobulins (TSI)	42	− 0·152

** highly significant

These correlations between first-degree relatives are merely the first step. For PBI and T_3 they would be compatible with a strong genetic component to the variation. But there is a long way to go before they can be attributed to genetic similarity. They need support from other degrees of relationship, and other factors also need to be considered. PBI levels for instance vary with age, sex, in females the position in the menstrual cycle, as well as with the most obvious factor of diet, and families of course share dietary habit. However, our series includes first-degree relatives living apart in different households, and there is no suggestion that these are less similar than those living in the same households. But certainly they indicate the variables to concentrate upon in the first instance in any more intensive investigation of the genetics of thyroid function.

The Threshold Model—Thyroid Antibodies

The polygenic model can be applied in those cases where a continuous distribution cannot be observed, but instead may be postulated so that the character can be regarded as occurring when the individual falls beyond a certain threshold (Falconer, 1965). For example, the population falls into two categories, those with thyroid autoantibodies and those without; but the liability to form thyroid autoantibodies does not behave as a simple Mendelian character, and its familial tendency requires a different explanation. This liability may be regarded as being a continuous variable differing from one individual to another, and having a normal distribution in the population. Those therefore who actually produce the antibodies when the appropriate stimulus is applied are those beyond a certain threshold in this distribution (Fig. 3). From the proportion of such individuals in the population, the distance of their mean liability from the population mean can be calculated; the mean distance of their first-degree relatives from the population mean would be half this were the character completely under genetic control, and from the deviation of the observed from the expected can be calculated the degree of genetic control.

From a large general practice near Newcastle upon Tyne, a 10 per cent sample was taken of all patients over the age of 60, selected at random from the practice list arranged in alphabetical order. It is in this age group, both in men and in women, that the incidence of antibodies, and hence presumably of expression of any underlying genes, is greatest (Dingle et al., 1966). In this series those patients with antibodies were taken as the propositi and those without were taken as normal controls. For each such proband or control, family information was collected relating to his children and his surviving parents and sibs. From each subject and each available first-degree relative blood was tested for thyroglobulin antibodies by the tanned red cell (TRC) technique. In all, specimens were tested from 322 individuals (Hall, Dingle and Roberts, 1972).

Of the 58 subjects, 14 (5 males, 9 females) were found to have positive thyroid antibodies at a titre of 1 in 25 and above, and these were taken as propositi. A further 8 (4 males, 4 females) showing a positive titre of 1 in 5 were discarded from the analysis. The remaining 36 showing a negative result were taken as controls. Relatives were regarded as positive only if their titre was 1 in 25 or greater. The

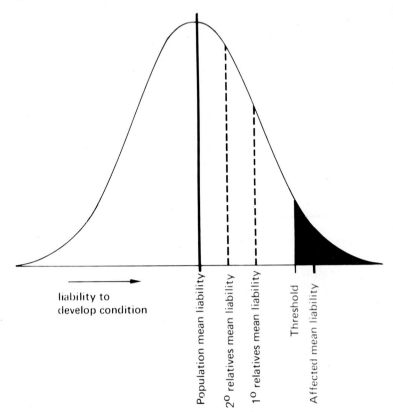

liability to
develop condition

Population mean liability

2⁰ relatives mean liability

1⁰ relatives mean liability

Threshold

Affected mean liability

FIG. 3.   Threshold model of genotypic liability.

results of the tests on first-degree relatives are shown in Table 2. The number of positives expected was calculated from the population incidence figures, standardizing the age distribution to that of the relatives. The number of positives observed in the relatives of propositi is, consistently, much higher than that expected from the population frequency (highly significantly so), and is significantly higher than in relatives of controls which resemble very closely those expected from the population figures (age-corrected). The increased incidence among first-degree relatives of propositi clearly suggests that heredity plays some part in the determination of the presence of antibodies. A segregation analysis in offspring from matings of different types gave, as expected, results that were not compatible with Mendelian single gene inheritance, because of the departure from the expected ratios in unions of different types.

TABLE 2.   Thyroid antibodies in first degree relatives.

Sex of subject	Sex of relative	No. tested	Observed No. positive	Observed % positive	Expected from population frequency % positive	Expected from population frequency No. positive
Affected						
Male	Male	4	1	25·0	10·7	0·4
	Female	7	3	42·9	16·4	1·1
Female	Male	15	2	13·3	4·3	0·6
	Female	27	8	29·6	16·5	4·5
			14			6·6
Control						
Male	Male	21	1	4·8	5·3	1·1
	Female	34	6	17·6	17·6	6·0
Female	Male	47	3	6·4	7·1	3·3
	Female	47	9	19·1	17·6	8·3
			19			18·7

Examining the data, however, on the hypothesis of polygenic control with a threshold distribution, the most reliable estimate derives from the comparison of the relatives of propositi with those of the controls. This gives an overall estimate of heritability of 53 ± 23 per cent. The genetic contribution to the liability to develop antibodies appears only moderately high. Again the investigation is relatively small, and it is necessary to extend it to relatives of more remote degree, for it is by comparing estimates from different degrees of relationship that the effects of other sources of resemblance can be excluded. Until this is done the present estimates should be regarded as upper limits.

Extended Family Analysis—Immunoglobulin Levels

For the most informative results from these two approaches, investigation needs to be extended to include second and third degree relatives as well as first degree for comparisons between these greatly facilitate interpretation. Sometimes however the nature of the population itself can be turned to advantage, to allow other informative ways of partitioning out the genetic contribution to the variance.

The circulating immunoglobulin levels represent the outcome of synthesis of these proteins in response to antigenic challenges. Their half-lives are relatively short, and so high circulating levels imply constant stimulus. Knowledge of the genetic control of these levels has lagged rather behind that of their structure in man. Some evidence has been forthcoming from the defective functioning of the immune system associated with a number of disease states in which a genetic component has been demonstrated, while twin studies of IgG, IgA and IgM have suggested "heritabilities" of 0·72, 0·79 and 0·83 respectively. Our study took the form of a family analysis. The families are Mohammedan Mandinko living in Keneba and Manduar, two villages in the Gambia, West Africa. Exposed as they are to the repeated assault of endemic and epidemic infection, there is ample environmental stimulus to immunoglobulin production. The villages have been the subject of detailed observation over a number of years, extensive and accurate data exist on the family relationships of the villagers, and their IgG and IgM levels were known to be high (Rowe et al., 1968).

The villages were surveyed three times, in March 1966, November 1966 and March 1967, and those individuals who were present at all three surveys and who were aged 2 years or more at the time of the first were included. For each individual the mean immunoglobulin level over all three surveys was calculated, and the individual's age taken at the midpoint between March 1966 and March 1967. Ig levels vary with age, and so age standardization was necessary and this was done on the assumption that the position of an individual in the distribution of Ig values at a given age remains constant; this assumption was subsequently justified by longitudinal study. The distribution of the age-standardized levels showed positive skewness with the longer tail of the distribution extending into the higher values, so all data were transformed to give a normalized distribution, and it is these transformed values that entered the family analysis.

The full results have been given elsewhere (Billewicz et al., 1974). Table 3 shows selected intrafamilial correlations. These are all positive but low, the midparent–child correlations tend to be slightly higher than between single parent and child and than the sib–sib correlations except for IgM, while the parent–parent correlations are quite compatible with zero values. These correlations indicate low to moderate heritabilities. From the full sib correlations maximum heritabilities for Ig G, A, M and D appear to be respectively 0·37, 0·61, 0·38 and 0·50; from the father–child regressions the heritabilities appear quite similar, 0·42, 0·60, 0·28 and 0·50; somewhat different, the midparent–child regression suggests heritabilities of 0·31, 0·57, 0·19, and 0·50, with the estimates for IgG and IgM here being rather lower.

TABLE 3.   Intrafamilial correlations of immunoglobulin levels.

	IgG	IgA	IgM	IgD
Parent–parent (91 pairs)	0·056 ± 0·104	−0·030 ± 0·105	−0·112 ± 0·104	0·098 ± 0·104
Father–child (103 fathers 308 offspring)	0·311	0·448	0·190	0·290
Mother–child (181 mothers 405 offspring)	0·248	0·316	0·195	0·242
Midparent– child (122 parent pairs 241 offspring)	0·317	0·543	0·185	0·389
Sib–sib (490 sibs, 178 sibships)	0·187	0·304	0·190	0·247

Here, however, the analysis is greatly facilitated by the occurrence of plural unions in these populations, as in many Islamic societies. This practice means there are many paternal half-sibs (i.e. children of the same father by different mothers). These are particularly useful for genetic analysis because offspring tend to be fed and generally cared for by the mother, so children of the same paternal genetic contribution are reared in different environments. It is therefore possible to partition the variance in the total offspring into a component attributable to differences between the children of different men; a component representing differences between children of different mothers by the same father; and a component attributable to differences among children by the one mother and father. These components of variation can be related to their underlying causal genetic and environmental sources, and so give a more accurate estimate of the heritability. In Table 4 the maternal component of variance is much greater than the paternal for IgA, M and D, indicating a substantial amount of variance due to maternal

TABLE 4.   Variance components.

	Between fathers	Between mothers within fathers	within mothers
IgG	0·074	0·036	0·825
A	0·047	0·305	0·718
M	0·058	0·169	0·755
D	0·096	0·194	0·714

environment. Heritability then is best estimated from the paternal component, and this gives heritability figures of 0·30, 0·18, 0·24, and 0·38 respectively. These are below the heritabilities estimated from the correlations and particularly for IgA. All immunoglobulin levels appear to be of low heritability. Hence the sib–sib and parent–offspring correlations owe much to common environment as well as to common genes. This example provides a useful reminder of some of the difficulties inherent in intrafamilial analyses of labile variables such as immunoglobulin levels and other physiological measures.

Pedigree Studies and the Effects of Inbreeding:
Electrocardiogram Analysis

For few physiological variables, apart from the monogenic clinical and threshold characters mentioned at the outset, is standard pedigree analysis likely to be rewarding, for continuously varying characters are not amenable to this approach. However, there is one genetic phenomenon which is known to affect a variety of continuously varying traits such as ages at which developmental milestones are reached, body measurements, neuromuscular tests, and school performance as shown by Schull and Neel (1965). This phenomenon is inbreeding, and for its analysis pedigree information is essential. Difficult to analyse in the United Kingdom population on account of the low frequency of consanguineous unions, its successful analysis demands particular populations characterized by appreciable frequencies of marriages between relatives, a wide range of inbreeding coefficients, and accurate pedigrees so that inbreeding can be accurately measured.

One such population is that of Tristan da Cunha, for which detailed pedigree information is available since its initial founding, where the development of inbreeding has been traced and where its effects on mental ability have already been indicated (Roberts, 1967). For the island population, ECG traces of excellent detail and quality are available, analysed by Professor Shillingford. Their genetic study incorporating a variety of measured parameters is progressing, but for the present illustration reference will be made to some of the results on 34 males and 38 females in whom the traces were completely normal. Measurements on all the ECG traces were made blind, i.e. without knowledge of any other detail of the individual concerned. The following intervals were measured on the lead II trace:

  (i) The PR interval, which measures the time required for the passage of the atrial depolarization wave to the vicinity of the atrioventricular node, the delay imposed by the slow-conducting junctional tissue surrounding this node, and the rapid spread of excitation via the atrioventricular bundle and its branches to the start of ventricular depolarization.

 (ii) The QT interval, which is the duration of the period between the initial negative deflection, signalling the onset of ventricular contraction, through the T wave indicating the repolarization of ventricular muscle which accompanies its relaxation.

(iii) The ST interval, which represents the time during which the muscle remains contracted.

The QT interval and the ST interval show no relationship with inbreeding (Table 5). However, the PR interval shows in both sexes a trend to diminution with increased inbreeding coefficient, and the regression is significant ($P < 0.02$) over the total sample. This trend is in the opposite direction to the effect on the PR interval of disease or drugs which depress conduction through the atrioventricular bundle, and implies increased conduction. One can only speculate on the reason for the association. It is not due to reduced pulse rate, for there is no suggestion of any effect of inbreeding on pulse rate in these subjects, and, if it were, one might expect the other intervals also to be affected. It may well be genetic; the depression is most

TABLE 5.   ECG intervals and inbreeding coefficients.

		Males		Females		Males + Females	
	F	*n*	mean	*n*	mean	*n*	mean
PR	0	7	0·177	6	0·160	13	0·169
	0·001–0·050	12	0·153	12	0·150	24	0·151
	0·051–0·100	11	0·155	17	0·145	28	0·149
	0·1	4	0·133	3	0·127	7	0·130
QT	0	7	0·359	6	0·368	13	0·363
	0·001–0·050	12	0·377	12	0·353	24	0·365
	0·051–0·100	11	0·361	17	0·361	28	0·361
	0·1	4	0·348	3	0·363	7	0·354
ST	0	7	0·276	6	0·290	13	0·282
	0·001–0·050	12	0·298	12	0·278	24	0·288
	0·051–0·100	11	0·280	17	0·289	28	0·285
	0·1	4	0·243	3	0·293	7	0·264

pronounced in those with inbreeding coefficients greater than 0·10, which suggests not only that recessive genes are involved but that these are distributed over a large number of loci. But certainly this suggestion calls for further investigation to establish whether or not a real trend exists and, if so, to enquire how it is brought about.

Epidemiological Studies and the Detection of Major Genes: Physiological Associations

If a physiological variable is under polygenic control, and if in the polygenic system there is also a major gene effect so that one allele appreciably increases the measurement by comparison with other alleles (as for example recently suggested for ABO blood group involvement in I.Q. by Gibson *et al.*, 1973), the distribution curves of those with and without the major gene would be laterally displaced. A simple comparison of the two distributions would indicate whether any measurable effect exists. To enquire whether such effects occurred, in the course of a local epidemiological survey covering nearly 3000 individuals (a one in six sample from the electoral role) from Whickham, a suburb of Newcastle, were examined some simple physiological measures and genetic polymorphisms. Against the phenotypes in nine blood group systems and nine red cell isoenzyme systems were examined the distributions of pulse rate, heart rate (from ECG), blood pressure systolic and diastolic, and the presence of hypertension; haemoglobin level and white cell count; fasting levels of serum cholesterol, triglyceride and glucose; and the presence or level of gastric parietal cell antibodies, thyopac-3, thyroglobulin antibodies, serum T_4, cytoplasmic thyroid antibodies, and other thyroid antibodies. The data were adjusted for age in those variables where an age-association was demonstrated.

The only association that proved significant in both sexes was between pulse rate and the Duffy blood group. There were fewer than expected Fy(b−) individuals with pulse rates of 70 or less and of 90 and over, and a higher than expected incidence of Fy(b−) phenotype in pulse rates 71–90. In this instance, although both sexes showed significant trends in the same direction, there is no clear gradient as would be expected on the polygenic model and the biological significance of the finding is dubious. The trends in the two sexes were in the same direction, but significant in only one, in heart rate from ECGs, which in dd individuals were lower than in rhesus positives, significantly so in females.

Two systems showed significant association with serum TSH levels. Males with phenotype Fy(a+) showed significantly higher mean levels than Fy(a−) in both uncorrected and age-compensated values, and likewise males of ADA 2–1 and 2–2 phenotypes showed significantly higher mean TSH levels in age compensated scores. The results were significant only in males, but in both cases small differences in the same direction were observed in females.

Fy(a−) phenotypes showed lower mean values of systolic blood pressure in females, which remained significant when corrected for age and fatness, while Fy(b+) phenotypes in males showed increased mean systolic blood pressures, and again the significance remained after age and adiposity correction.

The negative results are of interest in that they do not confirm the suggested relationship of serum cholesterol level with the ABO blood group and other phenotypes (Mayo et al., 1969, 1971; Medalie, 1970, 1971; Oliver et al., 1969; Singh and Orr, 1976) nor do they confirm the apparent association with the Lewis phenotypes (Beckman and Olivekrona, 1970; Langman et al., 1969).

In the Kidd blood group system also, a higher mean diastolic pressure occurred in females possessing the Kidd b gene (83·5 by comparison with 80·2 mm Hg) and the slight difference in males was in the same direction but the significance disappears on correcting for age. In no other marker system was there any parallel association in both sexes and significant in one.

Altogether 637 zero order statistical tests were performed, of which 36 gave significant results (30 at 5 per cent and 6 at 1 per cent, numbers very similar to those expected by chance association alone). These significant results are shown in Table 6. From the results so far it appears reasonable to conclude that none of the Mendelian marker systems examined is involved with major effect in the genetic control of these physiological responses.

Conclusion

The studies in the selection reported here are in varying degrees of completeness. No excuse is offered for the preliminary nature of several, for they have been chosen to illustrate not definitive findings but rather the range of methods available for seeking and elucidating genetic factors in physiological variation. They serve as a caution also, for though simple methods of heritability calculation, from threshold models or first-degree relative correlations, identify variates that are candidates for strong genetic control, extended familial analysis may indicate that resemblance between relatives owes more to common environment than to common genes. As regards the effects of major genes, while the negative results of the epidemiological approach may partly indicate that technical variation in field measurements of the variables may be too gross to allow the detection of any genetic element, this is less likely to apply to blood levels and similar measurements on specimens in modern laboratories which prove just as negative. It may well be however that against the enormous reservoir of genetical variation in the population the epidemiological approach is too gross, and that individual gene effects may only be detectable in finer intrafamilial analyses where background genetic "noise" is less, or in investigations designed to answer a specific question. For example, it is unlikely that the large Whickham survey would have identified the long-established association of ABO blood group with duodenal ulcer. Moreover, it would not have produced the suggestion that emerged from the present inbreeding study that recessive genes may be involved in control of heartbeat components, for this is pronounced especially in those individuals homozygous at more than 10 per cent of loci, and individuals as extreme as this are not likely to be encountered in an epidemiological survey in the United Kingdom. The point that emerges most clearly is the benefit to be derived from

TABLE 6. Significant associations of physiological variables with polymorphic systems.

Character	Marker system	Sex	χ^2	F	Level of significance	Degrees of freedom
Pulse rate	Fy^b	M	11·69	—	$P < 0.01$	2
Pulse rate	Fy^b	F	11·09	—	$P < 0.01$	2
Pulse (from ECG)	Rh D	F	—	4·36	$P < 0.05$	1
Blood pressure (diastolic)	Jk^a	F	—	8·21	$P < 0.01$	1
Blood pressure (diastolic)	6PGD	F	—	3·52	$P < 0.05$	2
Blood pressure (systolic)	6PGD	F	—	3·21	$P < 0.05$	2
Blood pressure (systolic)	PGM	M	—	2·70	$P < 0.05$	3
Hypertension	ABO	F	11·95	—	$P < 0.05$	4
Hypertension	Fy^a	F	7·54	—	$P < 0.01$	1
Hypertension	AP	M	11·19	—	$P < 0.05$	3
Haemoglobin level	MN	F	—	5·04	$P < 0.01$	2
White cell count	S	M	—	3·10	$P < 0.05$	2
White cell count	P	F	—	4·98	$P < 0.05$	1
Fasting cholesterol	S	M	—	3·39	$P < 0.05$	2
Fasting cholesterol	Fy^b	F	—	6·38	$P < 0.05$	1
Fasting cholesterol	6PGD	M	—	3·45	$P < 0.05$	3
Fasting cholesterol	PGM	F	—	6·03	$P < 0.01$	2
Fasting triglyceride	6PGD	M	—	2·67	$P < 0.05$	3
Fasting triglyceride	AK	F	—	4·07	$P < 0.05$	2
Fasting glucose	ABO	M	—	2·29	$P < 0.05$	5
Fasting glucose	S	M	—	3·56	$P < 0.05$	2
Gastric parietal cell antibodies	ABO	M	9·13	—	$P < 0.05$	2
Gastric parietal cell antibodies	MN	F	7·40	—	$P < 0.05$	2
Gastric parietal cell antibodies	S	F	7·36	—	$P < 0.05$	2
Gastric parietal cell antibodies	AP	M	8·87	—	$P < 0.05$	3
Thyopac 3	S	F	—	3·03	$P < 0.05$	2
Thyopac 3	Jk^a	F	—	4·91	$P < 0.05$	1
Thyopac 3	A	M	—	2·29	$P < 0.05$	5
Thyroglobulin antibodies	MN	M + F	6·35	—	$P < 0.05$	2
Serum TS	AK	F	—	3·25	$P < 0.05$	2
Serum T$_4$	MN	M	—	3·21	$P < 0.05$	2
Serum T$_4$	P	F	—	3·82	$P < 0.05$	2
Serum T$_4$	PGM	M	—	5·72	$P < 0.05$	1
Cytoplasmic thyroid antibodies	S	M	3·89	—	$P < 0.05$	1
Cytoplasmic thyroid antibodies	Fy^a	F	4·00	—	$P < 0.05$	1
Other thyroid antibodies	Jk^b	M	4·22	—	$P < 0.05$	1

choosing the population, or section of it, most appropriate to provide an answer to a given problem on account of its breeding pattern, preferably where the environment provides extreme stimulus to the character under investigation. The populations of mankind vary greatly in their social structures and mating patterns, and this variation has been virtually ignored in seeking solutions of genetic problems. Here perhaps is a useful lesson for physiological genetics.

Acknowledgments

Acknowledgment is gratefully made to Professor R. Hall, Dr. M. Tunbridge, Dr. D. C. Evered, and all the members of the Whickham field team and of the Department of Medicine, University of Newcastle, for the tolerance shown to our genetic studies over the years; to the stalwarts of the staff of the Department of Human Genetics, and in particular Dr. Papiha, Mr. Green and Miss Wastell for their help with laboratory analyses; and to Mrs. Vera Lamb and Dr. David Appleton for their help with the computations.

References

BECKMAN, L. and OLIVEKRONA, T. (1970) Serum cholesterol and ABO and Lewis blood groups. *Lancet*, **i**, 1000.

BILLEWICZ, W. Z., ROBERTS, D. F., McGREGOR, I. A., ROWE, D. S. and WILSON, R. J. M. (1974) Family studies in immunoglobulin levels. *Clin. exp. Immunol.*, **16**, 13–22.

The figures in Tables 3 and 4 of the present paper are corrected from those in the original publication, in which a computer programming error in calculating the degrees of freedom was subsequently detected.

DINGLE, P. R., FERGUSON, A., HORN, D. B., TUBMAN, J. and HALL, R. (1966) The incidence of thyroglobulin antibodies and thyroid enlargement in a general practice in North-East England. *Clin. exp. Immunol.*, **1**, 277–284.

FALCONER, D. S. (1965) The inheritance of liability to certain diseases estimated from the incidence among relatives. *Ann. hum. Genet.*, **29**, 51–76.

GIBSON, J. B., HARRISON, G. A., CLARKE, V. A. and HIORNS, R. W. (1973) IQ and blood groups. *Nature (Lond.)*, **246**, 498–500.

HALL, R., DINGLE, P. R. and ROBERTS, D. F. (1972) Thyroid antibodies: a study of first-degree relatives. *Clin. Genet.*, **3**, 319–324.

HARRISON, G. A. and OWEN, J. J. T. (1964) Studies on the inheritance of human skin colour. *Ann. Hum. Genet.*, **28**, 27–37.

KAHLON, D. P. S. (1973) Studies of human skin pigmentation, with special reference to Sikh and hybrid Indo–European families. Ph.D. thesis, University of Newcastle upon Tyne.

LANGMAN, M. J., ELLWOOD, D. C., FOOTE, J. and PYRIE, D. R. (1969) ABO and Lewis blood groups and serum cholesterol. *Lancet*, **ii**, 607.

MAYO, O., FRASER, G. R. and STAMATOYANNOPOULOS, G. (1969) Genetic influence on serum cholesterol in two Greek villages. *Human Heredity*, **19**, 86–99.

MAYO, O., WIESENFELD, S. L., STAMATOYANNOPOULOS, G. and FRASER, G. R. (1971) Genetical influences on serum cholesterol level. *Lancet*, **2**, 554–555.

MEDALIE, J. H., LEVINE, C., NEUFELD, H., RISS, E., DREYFUS, F., PAPIER, C., GOLD-BOURT, U., KAHN, H. and ORON, D. (1970) Blood groups, cholesterol and myocardial infarction. *Lancet*, **2**, 723.

MEDALIE, J. H., LEVINE, C., PAPIER, C., GOLDBOURT, U., DREYFUS, F., ORON, D., NEUFELD, H. and RISS, E. (1971) Blood groups and serum cholesterol among 10,000 adult males. *Atherosclerosis*, **14**, 219–229.

MOURANT, A. E., KOPEC, A. C. and DOMANIEWSKA-SOBCZAK, K. (1976) *The Distribution of the Human Blood Groups and Other Polymorphisms* (2nd edition). Oxford University Press.

OLIVER, M. F., GEIZEROVA, H., CUMMING, R. A. and HEADY, J. A. (1969) Serum cholesterol and ABO and Rhesus blood groups. *Lancet*, **2**, 605–606.

ROBERTS, D. F. (1952) An ecological approach to physical anthropology. *Proc. IVth Internat. Congr. Anthr. Sci.*, Vienna, 145–148. Verlag Adolf Holzhausens.

ROBERTS, D. F. (1967) Incest, inbreeding and mental abilities. *Br. Med. J.*, **4**, 336–337.

ROWE, D. S., McGREGOR, I. A., SMITH, S. J., HALL, P. and WILLIAMS, K. (1968) Plasma immunoglobulin concentrations in a West African (Gambian) community and in a group of healthy British adults. *Clin. exp. Immunol.*, **3**, 63.

SCHULL, W. J. and NEEL, J. V. (1965) *The Effects of Inbreeding on Japanese Children.* Harper & Rowe, New York.

SINGH, C. F. and ORR, J. D. (1976) Analysis of genetic and environmental sources of variation in serum cholesterol in Tecumseh, Michigan. III. Identification of genetic effects using 12 polymorphic genetic blood marker systems. *Am. J. Hum. Genet.*, **28**, 453–464.

TWIN STUDIES ON FUNCTIONAL CAPACITY

VASSILIS KLISSOURAS

Departments of Physiology and Physical Education, McGill University,
Montreal, Canada

Genetic Factors

THE functional capacity of an individual is the result of interaction between heredity and environment. However, wide inter-individual variability exists in adaptive responses, and one wonders to what extent individual differences are attributable to genetic variation and to what extent to environmental conditions. From a comparison of intra-pair differences between identical and non-identical twins, it is possible to answer this question since, in effect, phenotypic variability in identical twins is due solely to environmental agents, whereas that in non-identical twins is due to both genetic fluctuations and extragenetic influences.

In an early study based on the variance of such intra-pair twin differences we found that the contribution of heredity to the inter-individual differences in maximal oxygen uptake, which is used as a performance criterion of functional capacity, is relatively high (Klissouras, 1971). Fig. 1 depicts the data obtained in this study from 25 (15 MZ and 10 DZ) pairs of twin boys. The intra-pair difference tends to be smaller between identical than non-identical twins. In fact the experimental error could account for all intra-pair differences in identical twins. Statistical treatment revealed that the difference in the intra-pair variance between MZ and DZ twins was significant well beyond the 0·01 probability level. Young twins were used as subjects in this study to ensure that environmental influences were comparable for MZ and DZ twins. It could be argued, however,

43

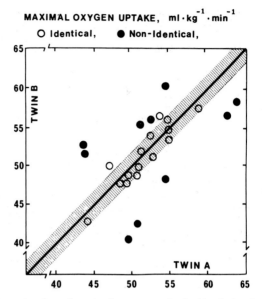

FIG. 1.   Intra-pair values of maximal oxygen uptake for identical and non-identical twin boys aged between 7 and 13 years (data from Klissouras, 1971).

that DZ pairs would be under more diverse environmental influences than MZ pairs during the developmental period. Thus, a follow-up study was conducted to determine whether the small intra-pair differences observed between identical twins and the marked differences between non-identical twins persist throughout life (Klissouras, Pirnay and Petit, 1973). Thirty-nine pairs of twins (23 MZ and 16 DZ) of both sexes, ranging in age from 9 to 52 years were used as subjects. In twins exposed to similar environments at different stages in their lives, any demonstrable differences between dizygotic as compared with monozygotic twins must be an expression of the relative strength of the genotype. In those exposed to contrasting environments, the resulting differences may provide a measure of this responsiveness to environmental forces. The results shown in Fig. 2 confirmed our earlier conclusion that heredity accounts almost entirely for existing differences in maximal oxygen uptake.

    One may still wonder whether intra-pair differences in maximal oxygen uptake could be related to differences in mode of life. Reference may be made to some case studies (Klissouras et al., 1973). In one case of non-identical twins aged 21 years who lived apart since 16, one twin had trained strenuously for competitive middle-distance running, whereas his brother had never participated in sports of any kind. It was therefore surprising to find that the untrained twin had a maximal oxygen uptake of $56.0 \, \text{ml} \, \text{kg}^{-1} \, \text{min}^{-1}$ as compared with a value of 52.8 for his trained counterpart. One cannot escape the inference that if it were not for the physical training the intra-pair difference between this twin pair would have been greater. Further, the implicit postulate of this observation is that some individuals with a weak genotype have to use a greater amount of physical activity to attain an average adaptive value, whereas those with generous native endowment may not need more than a threshold exposure to maintain their already high adaptive value. Another two cases are also intriguing. Two identical brothers, 40 years of age, had been separated at age 12 and had had different lifestyles. More important, one twin had engaged in vigorous train-ing for competitive basketball (12–30 years, 18–30 years on the national level), whereas his brother was only moderately active during the same period. For the last ten years neither of them had been involved in regular physical exertion. When tested, their

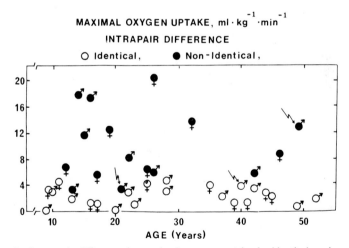

FIG. 2. Intra-pair difference in maximal oxygen uptake in identical and non-identical twins of different age. Arrows indicate three case studies discussed in the text (data from Klissouras et al., 1973).

maximal oxygen uptake was closely similar—the absolute values being 37·8 and 41·7 ml kg^{-1} min^{-1} for the trained and untrained twin, respectively. In another case, non-identical twins had a maximal oxygen uptake of 31·9 and 45·0 ml kg^{-1} min^{-1} at age 49. They had lived together all their lives, had the same profession and both played competitive soccer from early childhood until they were 22 years of age. These observations support the notion that natural tendency inevitably asserts itself.

A question of physiological importance is to define the biological attributes which are responsible for the production of the large differences in $\dot{V}_{O_2 max}$ between DZ twins and small differences between MZ twins. To do this we need to partition, in MZ and DZ twins, the oxygen transport and oxygen utilization systems, so that we can single out variables which play a determining role in setting the degree of intra-pair differences in maximal oxygen uptake. Howald analysed some ultrastructural components and the activities of some energy-transforming enzymes in the muscle cells of identical and non-identical twins (Baeriswyl, Luthi, Claasen, Moesch, Klissouras and Howald, 1977). Such a thorough and comprehensive study has not been yet conducted for the oxygen transport system. In this regard Fig. 3 shows that heart volume and mitochondrial volume are non-decisive factors.

Environmental Influences

The potency of environmental forces upon hereditary predisposition can be fully evaluated only if they are given a chance to act maximally. In this context, it is important to know the limits set by the genotype, the relative potency of training at different developmental ages and the extent to which genotype and training stimulus interact. These questions can be elucidated with the co-twin analysis, where each subject is accompanied by a genotypically identical control.

"Ceiling" of performance. All functional capacities and physiological processes in man, as in all species, have a genetically determined ceiling. For example, the upper limit of oxygen uptake is a little over 7 litres per min and that of cardiac output close to 40 litres per min. Additionally we find that ceilings characteristic of individual genotypes must exist at different levels (Bonnier and Hansson, 1948) and the question then arises as to what extent environmental

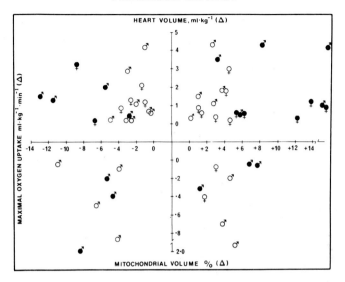

FIG. 3. Intra-pair differences in heart volume and percentage volume density of mitochondrial compared with corresponding intra-pair differences in maximal oxygen uptake in identical and non-identical twins of both sexes (heart volume data from Klissouras *et al.*, 1973; mitochondrial volume data from Baeriswyl *et al.*, 1977).

influences such as physical training can raise an individual's capacity above a certain level, towards the maximum value for the species.

To obtain some insight into this question, a pair of identical twins, a trained athlete and his untrained brother were tested over a period of $1\frac{1}{2}$ years. The untrained twin had a $\dot{V}_{O_2 max}$ of 35·9 ml kg^{-1} min^{-1}, whereas the trained twin had a peaked value of only 49·2 ml kg^{-1} min^{-1}. The latter value is comparable to an average maximum value of about 50 ml kg^{-1} min^{-1} for untrained college men of the same age, well below values reported for top athletes. So despite hard and prolonged training, the trained twin was unable to surpass an average level of adaptive capacity. The reason for this seems to hinge on his low pre-training functional adaptability as judged from that of his identical brother. This observation strongly suggests that rigorous athletic training cannot contribute to functional development beyond a limit set by the genotype. In this connection, the age-old question "Is an athlete born or made?" is meaningless as phrased. It is not a matter of predetermination versus plasticity, since heredity cannot operate in a vacuum and there must be an appropriate environment where the heredity factor attains full expression. What the question really endeavours to ask is "does everybody possess the constellation of genes or the genetic potential which, with appropriate training can find a phenotypic expression in superior athletic achievement?" In view of the empirical evidence, the answer to this question is unequivocally negative.

Early Training. There is much speculation but little evidence regarding the relative potency of training at different developmental ages. In a recent study (Weber, Kartodihardso and Klissouras, 1977), we split twelve pairs of identical twin boys (4 sets aged 10 years, 4 sets aged 13 years, and 4 sets aged 16 years), so that one twin trained, while his identical brother served as a control and continued in his normal day-to-day activity pattern. The training programme was of ten week duration and was designed to improve primarily the subject's endurance by both interval and continuous exercise. The individual differences for maximal oxygen uptake before and after training is shown in Fig. 4. The mean intra-pair difference was 11·5 per cent, and 13·9 per cent for the 10 and 16 year old groups, but the value was only 1·8 per cent in the 13 year olds. Since the type, intensity, duration and frequency of exercise were the same for all groups, the reason for this difference in response should be sought for in factors other than training. The most likely explanation for the commensurate increase in $\dot{V}_{O_2}$ in both trained and untrained 13 year-old twins, seems to hinge on the influences associated with the adolescent growth spurt that occurs at this age and is assessed by the height velocity. It is possible that hormonal activity is optimum during this age and any additional stimuli such as training cannot override its influence. In this connection one thinks of the anabolic activity of the growth hormone, which stimulates the transport of amino acids across cell membranes and the synthesis of protein. However, some other factors must play an essential role, since the blood growth hormone levels in children and adolescents are not different from that observed in adults during rest and in response to muscular work.

The question remains: At which developmental period is exercise most effective? Poupa and co-workers (1970) induced experimental cardiomegaly in rabbits and observed that animals in which the cardiomegaly was induced at an early age responded with an increase

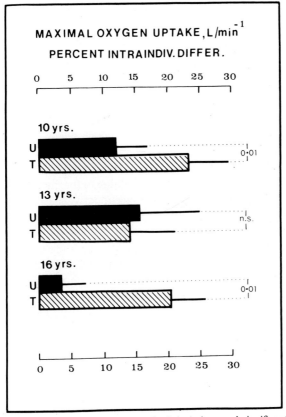

MAXIMAL OXYGEN UPTAKE, L/min^{-1}

PERCENT INTRAINDIV. DIFFER.

FIG. 4. Mean percentage differences, standard deviations and significant levels of maximal oxygen uptake for untrained (U) twins and their trained (T) brothers of three age groups (data from Weber *et al.*, 1977).

in the oxygen-consuming structures of the myocardium (cardiac cells) and the oxygen-supplying structures (terminal vascular bed per weight unit of cardiac tissue), whereas overloading of the heart during adulthood evoked development of the former structure, but not of the latter ones. They thus concluded that the ability of the heart to respond to the need for increased functional capacity is limited to the early post-natal period. The correspondent developmental period in man is not certain.

It is evident that the ontological time factor is decisive in development of functionally important structures. However, it remains uncertain at which developmental period the growth-promoting stimuli which act upon the tissues should be applied. The old hypothesis that more might be gained by introducing extra exercise at the time when the growth impulse is the strongest is not tenable any more in view of the present evidence.

Genotype-training interaction. A question of considerable theoretical and practical importance is whether different genotypes respond to a given training stimulus with a change of different magnitude. Split-twin experiments, in which one twin trains and his identical partner acts as a control, make it possible to separate the observed intra-pair variance into its three components: that due to heredity, that due to training and that due to the interaction between heredity and training. Eight twin boys underwent a 10-week training programme of the same amount and intensity, while their identical brothers restricted their activities to normal daily routines. The $\dot{V}_{O_2 max}$ of all twins was measused before and at the end of the 10-week period. The mean $\dot{V}_{O_2 max}$ for all experimental and control twins was $51 \cdot 9$ ml kg^{-1} min^{-1}, with non-significant intra-pair differences. The inter-pair variability ranged from $41 \cdot 1$ to $58 \cdot 6$ ml kg^{-1} min^{-1}, so that the interaction hypothesis could be tested. The mean $\dot{V}_{O_2 max}$ after training was $59 \cdot 4$ ml kg^{-1} min^{-1}, with adjustments for changes observed in the non-trained twins, and the range was $45 \cdot 2$ to $69 \cdot 3$ ml kg^{-1} min^{-1}. Treatment of the results by analysis of variance revealed that the interaction between genotype and training does not contribute significantly to the total variance (Table 1). These

TABLE 1.   Analysis of variance in $\dot{V}_{O_2 max}$. Figures based on eight twin pairs. The estimates of variances, in actual figures, are computed in the following way (n = number of twin pairs):

$$\text{Heredity} = \frac{\text{mean sq. Heredity} - \text{mean sq. interaction}}{2}$$

$$\text{Training} = \frac{\text{mean sq. training} - \text{mean sq. interaction}}{n}$$

Sources of variation	Mean squares	Variance in per cent of total variance
Training	221·72	42
Heredity	69·04	51
Interaction	4·39	7

findings do not support the notion that the magnitude of improvement in $\dot{V}_{O_2max}$ depends on the relative strength of the genotype. Thus, the inverse relationship occasionally observed between initial level of $\dot{V}_{O_2max}$ and relative improvement should be attributed to the amount and intensity of physical activity which presumably modifies the initial level of $\dot{V}_{O_2max}$. Further, it is surprising to find that in spite of strenuous training, the main cause of the total variance in $\dot{V}_{O_2max}$ is still the genetic predisposition. In this context, it should be pointed out that the partitioning of $\dot{V}_{O_2max}$ does not refer to the individual values but to the variation in a population. In view of the available evidence, it is concluded that variability in $\dot{V}_{O_2max}$ observed in a population that has been exposed to common environmental forces may be almost entirely determined by heredity, but its relative contribution to the total variance may be reduced to about 50 per cent with the operation of extreme environmental conditions.

Heritability Estimates

The extent to which genetic predisposition accounts for interindividual variation in a given attribute is commonly expressed by a heritability index. If the heritability index is unity, then the heredity may be considered the cause of the variation observed. If the coefficient is zero, the variation may be attributed solely to environmental influences. If the variation is partly affected by environment and partly conditioned by heredity, the index will fall between and its proximity to unity is taken as an indication of the relative strength of the genotype. Heritability indices are presented in Fig. 5 for individual differences in some metabolic, cardiorespiratory, pulmonary and neuromuscular attributes. It must be noted that the heritability estimates (H_{est}) were calculated only if the variance ratio ($F = W^2_{DZ}/W^2_{MZ}$) was significant in a statistical sense, for otherwise any inference drawn from these estimates would have little meaning. Several workers have disclosed similar results. Schwarz has reported in the Soviet literature H_{est} of 0·79 for $\dot{V}_{O_2max}$, 0·88 for PWC_{170} and 0·91 for anaerobic power (Schwartz, 1972), Kovár (1974) from Czechoslovakia has computed H_{est} for a number of variables related to motor performance. He found a H_{est} of 0·83 and 0·85 for ventical jump and throw of a medicine ball, both of which are tests of muscular power; the respective H_{est} for muscular force and endurance were 0·68 and 0·65. Venerando and Milani-Comparetti (1973)

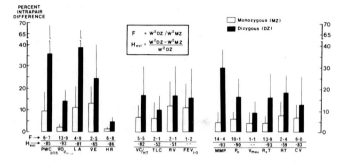

FIG. 5. Percentage intra-pair difference, their standard deviation, F-ratios and heritability estimates (H_{est}) of physical work capacity at a heart rate of 205 (PWC$_{250}$), maximal oxygen uptake ($\dot{V}_{O_2 max}$), maximal blood lactate concentration (LA), maximal work ventilation ($\dot{V}$E), maximal heart rate (HR), vital capacity to body height ratio (VC/Ht) total lung capacity (TLC), residual volume (RV), forced expiratory volume at one second (FEV 1·0''), maximal muscular power (MMP) of the forearm flexons, maximal isometric force (Po) of the forearm flexons, maximal speed ($\dot{V}_{max}$) of forearm flexons, reflex time ($R \times T$), reaction time (RT) and conduction velocity (CV). H_{est} were computed only when the F-ratio was significant at a level higher than 0·05 (based on data from Arkinstall et al., 1974, Klissouras, 1971, Klissouras et al., 1973, Komi et al., 1973, and Pirnay et al., 1972).

have conducted extensive twin studies in Italy and reported H_{est} for over 70 variables. Their H_{est} for maximal oxygen uptake (0·94), maximal work ventilation (0·83), maximal heart rate (0·91) and vital capacity (0·91) are comparable to our findings.

The validity of all these estimates depends upon the acceptability of the underlying assumptions. Two basic assumptions were made in the derivation of a H_{est}. It was assumed, first that no genetic–environment interaction is present and second, that environmental influences were comparable for MZ and DZ twins. In regard to the first assumption, we have obtained evidence to suggest that, the interaction between genotype and training does not contribute significantly to the total variance in maximal oxygen uptake, but we do not know whether such an interaction takes place for the other variables. In regard to the tenability of the second assumption, one has to consider both prenatal and postnatal environment. It has been argued that differences in intrauterine position and blood supply to the embryo, and accidental differences in the make-up of the cytoplasm may result in structural and biochemical differences between monozygous twins (Price, 1950). Although such differences in the prenatal environment may exist, no phenotypical differences that could be ascribed to their effect were observed in the identical twins studied. These observations strongly suggest either the equality of prenatal environment or that existing prenatal differences are not enduring, but are progressively equalized under the influence of a genetic maturational pacemaker (Sanform, Murrawski, Brazelton and Young, 1966). This, of course, would only apply to prenatal differences which do not result in injury of a vital organ, which in turn may cause some developmental anomaly or malformations. In any event, prenatal inequalities would only lead to an under-estimation of the share of heredity in the discordance of non-identical twins. As far as the postnatal environment is concerned it is believed that no differentiating influences can be operant if both MZ and DZ twins have similar lifestyles and if they are of a young age, for as children grow older the assumption of a shared environment becomes less certain. These H_{est} are based on twins of young age to ensure the environmental comparability. This does not mean that the environment did not vary, but that it varied approximately in the same direction and to the same degree for all individuals under study.

Finally, it must be noted that a high heritability index should not necessarily be interpreted to mean that the genetic factor has an etiologic role in the expressivity of the biological responses under study, nor has it sensible meaning with reference to measurement in an individual. It is erroneous, for example, to interpret a 93 per cent heritability index found for maximal aerobic power variance as meaning that 93 per cent of an individual's $\dot{V}_{O_2}$ is genetically deter-mined and 7 per cent susceptible to environmental modification. As Komi, Klissouras and Karvinen (1973) put it "The H_{est} is only an estimate of the extent to which heredity affects the variation of a given organic attribute, in a given population exposed to common environmental influences at a given time". Knowledge acquired from H_{est} should be complemented with information obtained from co-twin analysis on the potency of physical training and only then can the nature–nurture problem be placed in perspective and the environmental forces upon hereditary predisposition be evaluated.

Conclusion

In this paper I have tried to present some twin findings and inter-pret their meaning. These findings were derived from studies which have used the twin model and the co-twin analysis and were conducted either in my laboratory or in collaboration with Drs. Howald, Komi and Petit in their respective laboratories. In summary we found that:

(i) The genetic factor is the principal determinant of the variability in functional capacity as assessed by the maximal aerobic power observed among individuals regardless of age, who have lived under similar environmental conditions. Further, inter-individual differences in physical work capacity, anaerobic capacity, maximal force, reflex time and conduction velocity are also governed by genetic differences, while maximal work ventilation, total lung ventilation, residual volume, forced expiratory volume at 1·0 second, maximal speed of muscle shortening, and reflex time show almost as much diversity in DZ as in MZ twins.

(ii) The relative contribution of heredity to the total variance of functional capacity can be reduced to about 50 per cent with the operation of extreme environmental conditions. Habitual exercise can profoundly affect the expression of the genetic potential, but this can occur only within the fixed limits of heredity.

(iii) The ontological time factor may be decisive in development of functionally important structures, but the old hypothesis that more might be gained by introducing extra exercise at the time when the rate of growth is greatest is not tenable, and

(iv) Within the limitation of a narrow range of genetic variability observed in $\dot{V}_{O_2 \max}$, different genotypes respond to a given training stimulus with a change of the same magnitude.

Mention should also be made of some recent, as yet unpublished, observations by Komi and co-workers, according to which monozygous twin pairs have almost identical percentage distributions of the slow twitch fibres (ST %) in their vastus lateralis muscle, in contrast to dizygous twin pairs of both sexes. In view of the fact that a motor neuron innervates a both histochemically and physiologically uniform type of muscle fibre, the investigators suggested that the motor unit composition in human skeletal muscle is genetically fixed, and they inferred that the genetic factors are decisive in human potential performance and its prediction.

Acknowledgments

Our studies on twins have been supported by grants from the Medical Research Council of Canada and the Quebec Ministry of Education.

References

ARKINSTALL, W., NIRMEL, K., KLISSOURAS, V. and MILIC-EMILI, J. (1974) Genetic differences in the ventilatory response to inhaled CO_2. *J. appl. Physiol.*, **36**, 6–11.

BAERISWYL, C., LUTHI, Y., CLAASEN, H., MOESCH, H., KLISSOURAS, V. and HOWALD, H. (1977) Ultrastructure and biochemical function of skeletal muscle in twins: I. Genetical and environmental influences. *J. appl. Physiol.* (Submitted)

BONNIER, G. and HANSSON, A. (1948) Identical twin genetics in cattle. *Heredity*, **2**, 1–27.

KLISSOURAS, V. (1971) Heritability of adaptive variation. *J. appl. Physiol.*, **31**, 338–344.

KLISSOURAS, V., PIRNAY, F. and PETIT, J. M. (1973) Adaptation to maximal effort: Genetics and age. *J. appl. Physiol.*, **35**, 288–293.

KLISSOURAS, V. (1972) Genetic limit of functional adaptability. *Int. Z. Angew. Physiologie*, **30**, 85–94.

KLISSOURAS, V. (1973) Erblichkeit und Training-Studien mit Zwillingen. *Leistungsport*, **5**, 357–368.

KOMI, P., KLISSOURAS, V. and KARVINEN, E. (1973) Genetic variation in neuromuscular performance. *Int. Z. Angew. Physiol.*, **31**, 289–304.

KOVÁR, R. (1974) Prispevec Ke studiu Geneticke podminenosti Lidske motoriky. Thesis, Prague.

PIRNAY, J., KLISSOURAS, V. and PETIT, J. M. (1972) Fonction pulmonaire et génétique. *Acta Tuberc. Pneumol. Belg.*, **63**, 477–483.

POUPA, O., RALUSAN, K. and OSTADAL, B. (1970) *The Effect of Physical Activity upon the Heart of Vertebrates, in Physical Activity and Aging*. Ed. Brunner and Jokl. University Park Press, Baltimore.

PRICE, B. (1950) Primary biases of twin studies. *Am. J. Human Genet.*, **2**, 293–352.

SANFORM, G., MURRAWSKI, B. J., BRAZELTON, T. B. and YOUNG, G. C. (1966) Differences in individual development within a pair of identical twins. *Inter-Am. J. Psycho-Anal.*, **47**, 261–268.

SCHWARZ, V. (1972) On the relative role of genetic and environmental factors in development of physical work capacity in children: A twin-study (in Russian). Thesis, State University of Tartu, Soviet Union.

VENERANDO, A. and MILANI-COMPARETTI, M. (1973) Influenza dell'eredita sull attitudine ai vari Sport. *Medicina Dello Sport*, **26**, 347–352.

WEBER, G., KARTODIHARDSO, M. and KLISSOURAS, V. (1976) Growth and physical training with reference to heredity. *J. appl. Physiol.*, **40**, 211–215.

ULTRASTRUCTURE AND BIOCHEMICAL FUNCTION OF SKELETAL MUSCLE IN TWINS

H. Howald

Research Institute,
Swiss School for Physical Education and Sports,
Magglingen, Switzerland

Introduction

THE introduction of the muscle biopsy technique (Bergström, 1962) has made it possible in recent years to analyse the structure of the human skeletal muscle and its bioenergetic systems. There is ample evidence on the distribution of muscle fibre types (Gollnick, Armstrong, Saubert IV, Piehl and Saltin, 1972 a; Gollnick, Piehl, Saubert IV, Armstrong and Saltin, 1972 b; Schmalbruch, 1970), cellular ultrastructure (Hoppeler, Lüthi, Claassen, Weibel and Howald, 1973; Kiessling, Piehl and Lundquist, 1971; Morgan, Cobb, Short, Ross and Gunn, 1971), and activities of cellular enzymes (Gollnick *et al.*, 1972 a; Holloszy, 1965; Holloszy, Oscai, Don and Molé, 1970; Moesch and Howald, 1975; Pette and Staudte, 1971).

We studied ultrastructural components and the activities of some energy-transforming enzymes in the muscle cells of twins. We also made spirometric measurements in the same twins to find out whether cellular differences are related to differences in maximal oxygen uptake, which represents the global capacity of the organism to transport and utilize oxygen.

In the first part of our twin studies, we compared intra-pair differences of monozygotic and dizygotic twins to evaluate the genetic and non-genetic influences on structure and function of the muscle cell. The second part of the experiments was designed to

57

measure the influences of physical training as one single environmental factor in genetically identical subjects, namely monozygotic twins.

Subjects and Methods

Eleven pairs of monozygotic (MZ) and six pairs of dizygotic (DZ) twins volunteered for muscle biopsy and maximal exercise testing. The age and the anthropometric data of the twin groups are given in Table 1. Their zygosity was determined on the basis of a polysymptomatic comparison of morphological similarity and serological criteria (Giblet, 1969). A careful assessment of exposure to environmental differences, including medical as well as social history, was made for each twin set. In some twin pairs there were marked differences in physical activity.

TABLE 1. Age and anthropometric data of MZ and DZ twins. (x = mean, s_x = standard deviation.)

	MZ		DZ	
	$x \pm s_x$		$x \pm s_x$	
Age (years)	19·3	3·7	19·2	3·1
Weight (kg)	59·8	10·5	58·2	15·0
Height (cm)	168·8	9·3	169·7	10·6
Heart volume (ml/kg)	10·6	0·7	10·3	0·9

Two muscle samples (each 20–40 mg) were taken from the middle part of the vastus lateralis muscle with the needle biopsy technique (Bergström, 1962). Biopsies were taken early in the morning before ergometry. One sample of muscle tissue was used for electron microscopy. From each biopsy we cut six blocks and randomly took, from one oblique section of each, eight electron micrographs at a magnification of 75 000 (Hoppeler et al., 1973). Stereological analysis was performed with a 168-point short line test system (Weibel, 1973).

A second muscle sample was used for enzymatic analysis. It was immersed and homogenized manually in phosphate buffer. The activities of two mitochondrial (succinate-dehydrogenase, SDH, EC 1.3.99.1 and 3-hydroxyacyl-CoA-dehydrogenase, HAD, EC 1.1.1.35) and two extramitochondrial enzymes (hexokinase, HK, EC 2.7.1.1 and glyceraldehyde-3P-dehydrogenase, GAPDH, EC

1.2.1.12) were determined spectrophotometrically (Moesch et al., 1975).

Maximal oxygen uptake ($\dot{V}_{O_2max}$) was determined on a bicycle ergometer at progressively increasing intensity, using a fully electronic spirometer system with on-line computation of the data (Howald, 1973; Schönholzer, Bieler and Howald, 1973).

In the second part of our studies, on training, seven pairs of monozygotic twins (five male, two female, age 18 ± 3.4 years, range 15–25 years) volunteered for the training. Each pair of twins nominated one member to undergo a 23-week training while his identical brother or sister maintained his usual physical activity. The training programme consisted in running at 80 per cent of the individual maximum heart rate 3 times 15 minutes during the first week and was continuously increased to 3 times 30 minutes. This latter training duration was maintained from the tenth week to the end of the programme. Training effects were controlled monthly by Cooper's 12-minute test run and by monitoring the subject's submaximal heart rate at 80 per cent of the initial maximum work load on a Monark bicycle ergometer.

Results

Genetical influences

Mean intra-pair differences in cardiorespiratory responses to maximal effort are given in Table 2. The mean intra-pair variance computed for MZ twins was not statistically significant from that of DZ twins for all measurements, i.e. our MZ twins demonstrated as much diversity as DZ twins do. This finding seems to contradict the results of Klissouras, who found that heredity made a relatively high

TABLE 2. Analysis of cardiorespiratory responses to maximal work in monozygotic (MZ) and dizygotic (DZ) twins.

	Mean intra-pair difference and standard deviation	
	MZ	DZ
Work output (watts kg^{-1})	0.58 ± 0.29	0.57 ± 0.32
$\dot{V}_{O_2max}$ (ml min^{-1} kg^{-1})	5.0 ± 2.5	5.6 ± 2.5
Heart rate (beats min^{-1})	4.4 ± 4.4	7.7 ± 6.4
O_2 pulse (ml $beat^{-1}$ kg^{-1})	0.028 ± 0.013	0.022 ± 0.013

contribution to inter-individual differences in $\dot{V}_{O_2\,max}$ (Klissouras, 1971; Klissouras, Pirnay and Petit, 1973). However, in our subjects there were two pairs of MZ twins who had been exposed to contrasting environments. One pair of MZ twins had been separated for the last $1\frac{1}{2}$ years and had engaged in different activities very likely to produce intra-pair differences in $\dot{V}_{O_2\,max}$. One MZ twin had trained strenuously for competitive soccer, whereas his identical brother had never participated in sports, with the result that their intra-pair difference in $\dot{V}_{O_2\,max}$ reached 28 per cent and was the largest observed. Moreover, in our DZ twins there was one pair showing a very low intra-pair difference in $\dot{V}_{O_2\,max}$: one subject had a $\dot{V}_{O_2\,max}$ of $2 \cdot 6\,l\,min^{-1}$ which gave a relatively high figure of $64 \cdot 1\,ml\,min^{-1}\,kg^{-1}$ when related to his very low body weight. His non-identical brother attained a $\dot{V}_{O_2\,max}$ of $3 \cdot 7\,l\,min^{-1}$, or $62 \cdot 9\,ml\,min^{-1}\,kg^{-1}$ and was much more active in sports. Thus the intra-pair difference in $\dot{V}_{O_2\,max}$, which otherwise could have been expected because of constitutional and genetical influences, had probably been reduced. If the $\dot{V}_{O_2\,max}$ of these three twin pairs are excluded from the computation to satisfy the assumption of comparability of environmental influences, then there is a significant difference ($2P < 0 \cdot 02$) between the MZ and DZ groups and a heritability index of 68 per cent (Klissouras, 1971).

Average intra-pair differences calculated for MZ and DZ twins for the ultra-structural parameters are shown in Table 3. Again, the intergroup differences were not statistically significant, neither for volume densities nor for surface densities of mitochondria.

TABLE 3.   Analysis of morphometric differences between monozygotic (MZ) and dizygotic (DZ) twins.

	Mean intra-pair difference and standard deviation	
	MZ	DZ
Volume density of mitochondria (%)	$0 \cdot 46 \pm 0 \cdot 37$	$0 \cdot 32 \pm 0 \cdot 22$
Mitochondrial volume to myofibril volume ratio ($\times 10^{-2}$)	$0 \cdot 64 \pm 0 \cdot 59$	$0 \cdot 56 \pm 0 \cdot 70$
Surface density of mitochondrial inner membranes ($m^2\,cm^{-3}$)	$0 \cdot 125 \pm 0 \cdot 088$	$0 \cdot 110 \pm 0 \cdot 088$
Surface density of mitochondrial outer membrane ($m^2\,cm^{-3}$)	$0 \cdot 035 \pm 0 \cdot 025$	$0 \cdot 057 \pm 0 \cdot 061$

The mean intra-pair differences of the enzymatic activities for the MZ and DZ twins are shown in Table 4. The statistical comparison of intra-pair variances revealed that MZ twins demonstrated as much diversity as did DZ twins in the activities of HK and SDH. A significant intra-pair difference was observed between the MZ and DZ twins only in the activity of GAPDH and HAD. In order to combine structure and biochemical function of those enzymes that are bound to mitochondrial inner membranes, we computed correlation coefficients and found that SDH and HAD activities are related significantly to mitochondrial volume density. The individual intra-pair differences in $\dot{V}_{O_2 max}$ were not parallel with proportional differences either in morphometric parameters or in enzymatic activities. How then is the intra-pair variability in $\dot{V}_{O_2 max}$ explained? This is an important question that cannot be resolved at the present time. Our data merely indicate the varying importance of both, the oxygen transport and the oxygen utilization system in setting the degree of intra-pair differences.

TABLE 4. Analysis of the difference in muscular enzymatic activity in monozygotic (MZ) and dizygotic (DZ) twins.

	Mean intra-pair difference and standard deviation	
(mmol min^{-1} kg^{-1})	MZ	DZ
Hexokinase (HK)	0.23 ± 0.24	0.43 ± 0.15
Glyceraldehyde-3P-dehydrogenase (GAPDH)	37.0 ± 16.96	$139.0 \pm 56.57*$
Succinate-dehydrogenase (SDH)	0.77 ± 0.67	0.49 ± 0.36
3-Hydroxyacyl-CoA-dehydrogenase (HAD)	1.76 ± 1.60	$4.10 \pm 1.13**$

$* P < 0.05; ** P < 0.01.$

Training influences

The second part of our studies was designed to investigate the ultrastructural and biochemical adaptations taking place in the human skeletal muscle by an endurance training programme of relatively low intensity. In order to exclude genetic and inter-individual influences as much as possible, we examined identical twins longitudinally.

After the training period, the physiological, ultrastructural, and biochemical parameters of the trained subjects and of the control group were determined as before. For both groups, the individual differences between the two examinations were calculated. Assuming the differences in the non-training group to be due to different factors (e.g. seasonal influences, experimental procedure, growth etc.), the variation between the intergroup differences ($\Delta_{\Delta_T - \Delta_U}$) can be fully attributed to the training stimulus.

The training-induced variations in anthropometric, respiratory, and cardiovascular parameters are shown in Fig. 1. The columns represent the mean percentage variation before and after training. Confidence levels of statistical significance are given at the bottom of the figure. The 23-week running programme caused significant increases in maximum ergometric performance (10·1 per cent), heart volume (7·5 per cent), $\dot{V}_{O_2 max}$ (15·4 per cent), and maximum oxygen pulse (15·6 per cent). The decreases in heart rate at our first ($-6·9$ per cent) and second ($-4·7$ per cent) submaximal work load as well as the 12·2 per cent increase in maximum pulmonary ventilation were different at the 10 per cent confidence level of statistical significance, whereas the decreases in body weight ($-1·9$ per cent), adiposity ($-10·2$ per cent), and maximum heart rate ($-0·8$ per cent) were statistically not significant.

Figure 2 shows the ultrastructural adaptations in m. vastus lateralis produced by training. There was a statistically significant

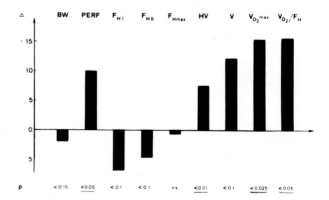

FIG. 1.   Mean percentage variation and statistical significance of changes in anthropometric, respiratory, and cardiovascular parameters due to training (BW = body weight, PERF = maximum ergometric performance, F_{HI} = heart rate at first submaximal work load, F_{HII} = heart rate at second submaximal work load, $F_{H max}$ = maximum heart rate, HV = heart volume, $\dot{V}$ = maximum ventilation, $\dot{V}_{O_2 max}$ = maximum oxygen uptake, $\dot{V}_{O_2}/F_H$ = oxygen pulse).

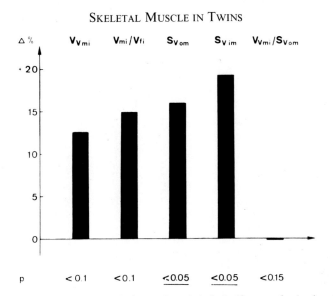

FIG. 2. Mean percentage variation and statistical significance of mitochondrial adaptation to training ($V_{V_{mi}}$ = mitochondrial volume density, V_{mi}/V_{fi} = mitochondrial to myofibrillar volume ratio, $S_{V_{om}}$ = surface density of mitochondrial outer membrane, $S_{V_{im}}$ = surface density of mitochondrial inner membranes. $V_{V_{mi}}/S_{V_{om}}$ = volume to surface ratio of mitochondria).

increase in the surface densities of mitochondrial outer (16·0 per cent) and inner membranes (19·3 per cent). The increases in volume density of mitochondria (12·6 per cent) and in the mitochondrial to myofibril volume ratio (14·9 per cent) were different at the 10 per cent confidence level of statistical significance. The decrease in volume to surface ratio of mitochondria (0·03 per cent) was statistically not significant.

The changes in the activities of extra- and intramitochondrial enzymes are shown in Fig. 3. Training produced a statistically significant increase in the activity of the intramitochondrial enzyme HAD (23·2 per cent) and in the one of the partly intramitochondrial enzyme malate-dehydrogenase, MDH (32·6 per cent). The 28·1 per cent increase in the activity of SDH, which is bound to mitochondrial inner membranes, and the 17·4 per cent increase in the activity of the extramitochondrial enzyme HK were statistically not quite significant. For SDH, the mean pre- to post-training difference was not statistically significant because one of the non-training control subjects showed a higher SDH activity in his second muscle biopsy.

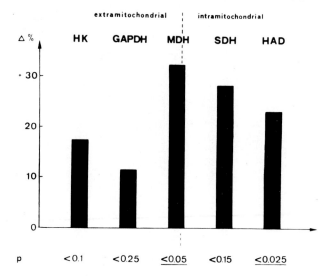

FIG. 3. Mean percentage variation and statistical significance of changes in extra- and intra-mitochondrial enzyme activities (HK = hexokinase, GAPDH = glyceraldehyde-3P-dehydrogenase, MDH = malate-dehydrogenase, SDH = succinate- dehydrogenase, HAD = 3-hydroxyacyl-CoA-dehydrogenase).

Excluding this single value from the calculation, the change in mean activity of SDH would be highly significant ($P < 0.01$). For the activity of GAPDH, an enzyme involved in the anaerobic glycolysis, there was no significant change in response to endurance training.

Compared to the intensity and duration of modern endurance training in athletes, the running programme used in this study was rather minimal. Nevertheless, it induced all the respiratory, cardiovascular, ultrastructural, and biochemical adaptations which have been found in well-trained athletes (Hoppeler *et al.*, 1973, Moesch *et al.*, 1975, Schönholzer *et al.*, 1973). Since some of the changes induced were not highly significant, it appears that the amount of training used in the present study is perhaps the smallest which can cause physiological differences to occur in the human oxygen transport and utilization systems.

Discussion

To summarize our twin studies, it may be concluded that as far as physical performance capacity, cardiorespiratory adaptation and ultrastructural as well as biochemical features of the skeletal muscle are concerned, environmental influences, and especially physical activity, are stronger than the genetical background.

Acknowledgments

This study was supported by grants from Schweizerischer Nationalfonds zur Förderung der wissenschaftlichen Forschung. The author thanks the subjects for their cooperation throughout the study, and Claude Baeriswyl, Yvonne Lüthi, Hugo Moesch, Hans Spring, Helgard Claassen, Ruth Dienel, Theres Appenzeller, and Erwin Leiser for their technical help. Appreciation is also expressed to Rose-Marie Engel for typing the manuscript.

References

BERGSTRÖM, J. (1962) Muscle electrolytes in man. *Scandinavian Journal of Clinical and Laboratory Investigation*, **14**, suppl. 68.

GIBLET, E. R. (1969) *Genetic Markers in Human Blood.* Oxford: Blackwell Scientific Publications.

GOLLNICK, P. D., ARMSTRONG, R. B., SAUBERT IV, C. W., PIEHL, K. and SALTIN, B. (1972 a) Enzyme activity and fiber composition in skeletal muscle of trained and untrained men. *Journal of Applied Physiology*, **33**, 312–319.

GOLLNICK, P. D., PIEHL, K., SAUBERT IV, C. W., ARMSTRONG, R. B. and SALTIN, B. (1972 b). Diet, exercise and glycogen changes in human skeletal muscle fibers. *Journal of Applied Physiology*, **33**, 421–425.

HOLLOSZY, J. O. (1967) Biochemical adaptations in muscle. Effects in exercise on mitochondrial oxygen uptake and respiratory enzyme activity in skeletal muscle. *Journal of Biological Chemistry*, **242**, 2278–2282.

HOLLOSZY, J. O., OSCAI, L. B., DON, I. J. and MOLE, P. A. (1970) Mitochondrial citric acid cycle and related enzymes: adaptive response to exercise. *Biochemical and Biophysical Research Communications*, **40**, 1368–1373.

HOPPELER, H., LÜTHI, P., CLAASSEN, H., WEIBEL, E. R. and HOWALD, H. (1973) The ultrastructure of the normal human skeletal muscle. A morphometric analysis on untrained men, women and well-trained orienteers. *Pflügers Archiv*, **344**, 217–232.

HOWALD, H. (1973) Eine Ergospirometrie-Anlage mit on-line-Datenverarbeitung durch Mikro-computer. *Acta Medicotechnica*, **21**, 115–120.

KIESSLING, K. H., PIEHL, K. and LUNDQUIST, C. G. (1971) Effect of physical training on ultrastructural features in human skeletal muscle. I, *Muscle Metabolism during Exercise*, ed. Pernow, B., and Saltin, B., p. 97. New York, London: Plenum Press.

KLISSOURAS, V. (1971) Heritability of adaptive variation. *Journal of Applied Physiology*, **31**, 338–344.

KLISSOURAS, V., PIRNAY, F. and PETIT, J. M. (1973) Adaptation to maximal effort: Genetics and age. *Journal of Applied Physiology*, **35**, 288–293.

MOESCH, H. and HOWALD, H. (1975) Hexokinase (HK), glyceraldehyde-3P-dehydrogenase (GAPDH), succinate-dehydrogenase (SDH), and 3-hydroxyacyl-CoA-dehydrogenase (HAD) in skeletal muscle of trained and untrained men. In *Metabolic Adaptation in Prolonged Physical Exercise*, ed. Howald, H. and Poortmans, J. R., p. 463. Basel: Birkhäuser Verlag.

MORGAN, T. E., COBB, L. A., SHORT, F. A., ROSS, R. and GUNN, D. R. (1971) Effects of long-term exercise on human muscle mitochondria. In *Muscle Metabolism during Exercise*, ed. Pernow, B. and Saltin, B., p. 87. New York, London: Plenum Press.

PETTE, D. and STAUDTE, H. W. (1971) Differences between red and white muscles. In *Limiting Factors of Physical Performance*, ed. Keul, J., p. 23. Stuttgart: G. Thieme Publishers.

SCHMALBRUCH, H. (1970) Die quergestreiften Muskelfasern des Menschen. *Ergebnisse Anatomie und Entwicklungsgeschichte*, **43**, 1.

SCHÖNHOLZER, G., BIELER, G. and HOWALD, H. (1973) Ergometrische Methoden zur Messung der aeroben und anaeroben Kapazität. In *III. Internationales Seminar für Ergometrie*, ed. Hansen, G. and Mellerowicz, H., p. 84. Berlin: Ergon Verlag.

WEIBEL, F. R. (1973) Stereological techniques for electron microscopic morphometry. In *Principles and Techniques of Electron Microscopy*, ed. Hayat, M. A., volume 3, p. 237. New York: Van Nostrand Reinhold Company.

CHEMICAL CONTROL OF BREATHING IN IDENTICAL TWIN ATHLETES

A. G. Leitch

Department of Medicine, Royal Infirmary, Edinburgh

Introduction

THE contribution of genetic and environmental factors to physiological variables has always been difficult to quantitate. Studies of identical twins have shown a close correlation between twins in vital capacity (Arkinstall, Nirmel, Klissouras and Milic-Emili, 1974) and maximal oxygen uptake (Klissouras, Pirnay and Petit, 1973). No such correlation was found for measurements of the ventilatory response to carbon dioxide (Arkinstall *et al.*, 1974) and the ventilatory response to hypoxia has never been reported in identical twin subjects.

The degree of physical activity is one environmental factor which is difficult to quantify in any identical twin study. It is known that decreased (Saltin, Blomquist, Mitchell, Johnson, Widdenthal and Chapman, 1968) and increased (Astrand and Rodahl, 1970) physical activity can produce marked changes in maximal oxygen uptake. It has also been shown that there may be a relationship between the ventilatory response to carbon dioxide and different types of athletic activity (Rebuck and Read, 1970). I have been able to eliminate variations in physical activity from the environmental factors affecting maximal oxygen uptake and chemical control of breathing, by studying identical twin athletes who were in training for the same events.

Subjects and Methods

The subjects A and B, were two 16-year-old Scottish female twins who were identical in physical features and therefore almost certainly monozygotic (Nichols, 1965). In addition, the HL–A phenotype and all erythrocyte antigens tested (ABO, MNS_s, P, Kk, Le^a, Fy^a, Kp^b, Wr^a, C^w and Rh) were identical. Each was in training for the Scottish National Team and each had represented their country at junior level. Both participated in the 400 and 800 m track events, but subject A was better at the 400 m distance and subject B performed better over 800 m. Each is currently ranked in the top five for these events in Scotland. Heights and weights of the subject are shown in the table with the best times for their chosen events (1974).

TABLE 1.   Height, weight and results obtained for identical twins A and B.

	Twin A	Twin B	Normal range
Height (cm)	161	161	
Weight (kg)	51·0	49·5	
Haemoglobin (g/100 ml)	13·7	13·1	12·0–16·0
Time for 400 m (s)	55·0	—	
Time for 800 m (s)	—	129·3	
$\dot{V}_{O_2 max}$ (ml min^{-1} kg^{-1})	59	60	53–58
Ventilatory response to CO_2 (rebreathing)			
(a) sCO_2 (l min^{-1} kPa^{-1})	8·4	5·0	4·3–61·3
(b) frequency (breaths/min) at pCO_2 8 kPa	16·6	18·3	
pCO_2 8·67 kPa	18·8	18·7	
Hypoxic index (%)	19	22	13–112
Lung volumes (litres)			
Functional residual capacity	2·91	2·91	
Vital capacity	3·65	3·50	
Expiratory reserve volume	1·55	1·55	
Residual volume (RV)	1·36	1·36	
Total lung capacity (TLC)	5·09	5·16	
RV/TLC (%)	28	26	
Forced expiratory volume 1 s (FEV)	3·45	3·70	
Forced vital capacity (FVC)	3·80	3·90	
FEV/FVC (%)	91	95	

$\dot{V}_{O_2 max}$ = maximal oxygen uptake; sCO_2 = slope of the line relating ventilation (l/min BTPS) to end-tidal pCO_2.

Lung volumes were measured by helium dilution and maximal oxygen uptake ($\dot{V}_{O_2 max}$) by a modification of the method of Taylor, Buskirk and Henschel (1955). After a "warming-up" run at 11·25 km h^{-1} (7 m.p.h.) on the level treadmill for 8 min, breathing through an Otis–McKerrow valve, the subjects ran at the same speed up a 5° gradient for 5 min, expired air being collected for measurements of

O_2 and CO_2 concentrations and minute ventilation being measured during the last minute. After a 10 min rest the subjects then ran for 3 min up a 7° slope, measurements again being made in the last minute of exercise. Neither subject could sustain this exercise for 6 min, thus indicating that $\dot{V}_{O_2 \, max}$ had been attained (Mitchell and Blomquist, 1971). This was confirmed by measurements during the last minute of a run for $2\frac{3}{4}$ min up an 8° slope. Values for $\dot{V}_{O_2}$ obtained in this final run did not differ by more than 30 ml/min from the preceding ones, thereby confirming that $\dot{V}_{O_2 \, max}$ had been attained. Inspiratory and expiratory resistance was 0·098 kPa (1 cm H_2O) at 1·5 l/s.

The ventilatory response to CO_2 was measured with a modification of Read's (1967) rebreathing technique. The subject rebreathed from a 6 l bag containing $O_2 + CO_2$ (93:7) via a valve and pneumotachograph (Fleisch No. 3). A Varian M3 mass spectrometer probe monitored pO_2 and pCO_2 in the rebreathing system. Volume was calibrated before and after each study, the gas mixture remaining in the bag at the end of rebreathing being used, and the mass spectrometer was calibrated at this time with gas mixtures analysed by the Lloyd–Haldane apparatus. Off-line computer analysis from magnetic tape recordings yielded values of frequency, tidal volume, instantaneous minute ventilation, pO_2 and pCO_2 for each breath. The slope of the CO_2 response curve was calculated from the linear regression of expired minute ventilation on pCO_2, after omitting the first 30 s of the record. Duplicate measurements were performed in each subject, the mean value being taken for each pO_2 always remained above 26·7 kPa (200 mm Hg) in the rebreathing study.

The hypoxic drive to breathing was assessed by the method of Flenley, Cooke, King, Leitch and Brash (1973), which utilizes the known potentiation of this drive by exercise (Asmussen and Neilsen, 1947). The subjects exercised on a level treadmill, attaining a steady-state $\dot{V}_{O_2}$ of 950 ml/min when breathing air, and were exposed at approximately 5 min intervals to a hypoxic stimulus in the form of three breaths of 100 per cent nitrogen. The method of expressing the ventilatory response to this hypoxic stimulus as a hypoxic index, the apparatus used and the normal range of response in healthy men have been previously described (Flenley et al., 1973).

The hypoxic drive to breathing was also assessed at rest and on exercise ($\dot{V}_{O_2} = 1·0$ l/min) during isocapnic progressive hypoxia as described by Weil, Byrne-Quinn, Sodal, Friesen, Underhill, Filley and Grover (1970). The pO_2 was lowered from 13·3 kPa (100 mm Hg) to 5·3 kPa (40 mm Hg) over 15 min while the pCO_2 was maintained constant at resting values by the addition of CO_2 to the inspired air. The pCO_2 did not vary by more than 0·27 kPa (2 mm Hg) during these studies.

The steady state ventilatory response to CO_2 was also assessed in both twins at four different levels of pO_2 (28, 9·33, 6·67 and 5·33 kPa) using the method of Cunningham, Shaw, Lahiri and Lloyd (1961), measurements of pCO_2, pO_2 and ventilation being made in the last three minutes of 10 min in the steady state using apparatus previously described (Flenley et al., 1973) and on-line computer analysis with a PDP 11–40 computer.

Results

The table shows that the $\dot{V}_{O_2\,max}$, lung volumes and ventilatory responses to CO_2 (rebreathing) and hypoxia (transient) are identical for each twin within the limitations of the methods employed. The normal range for ventilatory response to CO_2 is taken from the work of Rebuck and Read (1971) and our own studies on athletes, for $\dot{V}_{O_2max}$ from Saltin and Astrand's (1967) on Swedish national athletes (400–800 m events for women) and for transient hypoxia from the studies of Flenley et al. (1973) on healthy males and from studies on male and female Scottish athletes (A. G. Leitch and L. Clancy, unpublished work).

Figures 1 and 2 show the progressive hypoxia results at rest in the two twins. Each point represents simultaneous pO_2 and instantaneous minute ventilation measurements. The superimposed lines represent the lowest, mean and highest responses found in Hirschman, McCullough and Weil's (1975) study of 44 normal subjects. The lowest response is typical of the mean response in athletic subjects (Byrne-Quinn, Weil, Sodal, Filley and Grover, 1971) and is similar to that found in our twin athletic subjects. Figures 3 and 4 show the same responses on exercise with a $\dot{V}_{O_2}$ of 1000 ml/min in the twins with the mean normal value taken from the work of Weil, Byrne-Quinn, Sodal, Kline, McCullough and Filley (1972). The twin's responses are again identical but below the mean normal response.

Figure 5 shows the steady-state response to inhaled CO_2 at four

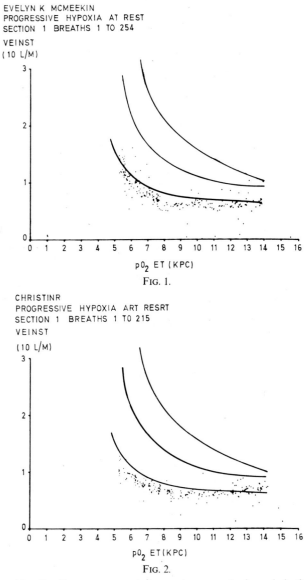

EVELYN K MCMEEKIN
PROGRESSIVE HYPOXIA AT REST
SECTION 1 BREATHS 1 TO 254

VEINST
(10 L/M)

pO_2 ET (KPC)

FIG. 1.

CHRISTINR
PROGRESSIVE HYPOXIA ART RESRT
SECTION 1 BREATHS 1 TO 215

VEINST
(10 L/M)

pO_2 ET(KPC)

FIG. 2.

FIGS. 1 and 2. Ventilatory responses to isocapnic progressive hypoxia in the twins at rest. The superimposed lines represent the highest, mean and lowest normal relationship between $\dot{V}_E$ and pO_2 (see text). Each point represents the instantaneous minute ventilation and pO_2 for a single breath.

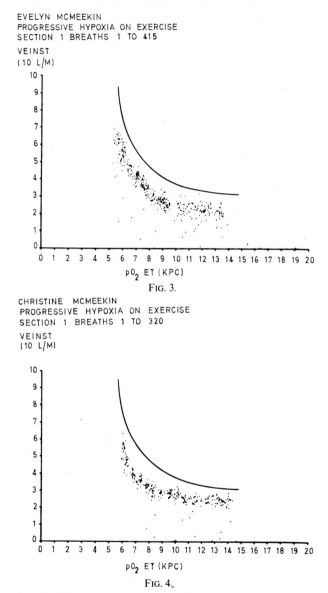

Figs. 3 and 4.   Ventilatory responses to isocapnic progressive hypoxia in the twins on exercise. The superimposed line represents the mean normal relationship between ventilation and pO_2 (see text).

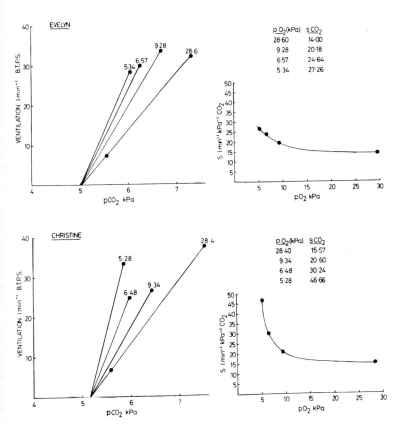

FIG. 5. The steady state ventilatory responses to CO_2 at different levels of pO_2 in the twins. The figures on the left show this relationship with the pO_2 at which each line was measured next to the line. On the right the slope of the line is plotted against the pO_2 at which it was measured. Inset are the values for the slope (sCO_2) of the line and the pO_2 at which it was measured.

levels of pO_2 for the two twins as the relationship between ventilation and pCO_2 on the left and as the relationship between the slope of the $\dot{V}_e / pCO_2$ line and the pO_2 at which it was measured on the right. At pO_2 of 28 kPa and 9·33 kPa the twins were identical within the limitations of the method but at lower pO_2's (6·67 and 5·33 kPa) larger differences were seen.

Discussion

The $\dot{V}_{O_2max}$ results compared favourably with values reported for Swedish national athletes (Saltin and Astrand, 1967) and are consistent with the finding by Klissouras et al. (1973) in 23 monozygotic twins of an insignificant mean intra-pair difference for $\dot{V}_{O_2max}$. The full lung volumes recorded in the subjects of this study are identical within the limitations of the method and expand the finding by Arkinstall et al. (1974) of identical vital capacities in 17 monozygotic twins.

Arkinstall et al. (1974) were unable to demonstrate identity of ventilatory response to CO_2 in their monozygotic twins, there being no difference in the intra-pair variance between their monozygous and dizygous twins. They attributed this to difference between twins in their frequency and tidal volume response to inhaled CO_2, the frequency response being determined by personality factors whereas there is a strong genetic component to the tidal volume response. The frequency response in the present study for both twins were similar (Table 1) during rebreathing and steady state and this would explain the similarity of their CO_2 responses. I did not assess personality but feel the physical training was probably a more important determinant of the frequency response in these two twins than any other factor.

Assessment of the hypoxic drive to breathing by measuring the ventilatory responses to transient hypoxia on exercise and progressive hypoxia at rest and on exercise shows that the responses are identical within the limitations of the methods in the two twins. In steady-state, however, their responses are identical only in hyperoxia and moderate hypoxia (pO_2 9·33 kPa), differences appearing with more severe hypoxia. In view of the demonstration of similar responses to hypoxia with the other methods, it is tempting to speculate that the differences observed in severe steady-state hypoxia do not represent differences in carotid body response but rather differences in the central nervous system response to hypoxia which is known to influence ventilation, particularly in steady-state studies (Kronenberg, Hamilton, Gabel, Hichey, Read and Severinghaus, 1972).

Although it is obviously impossible to draw firm conslusions from a study of one pair of identical twins, I feel that these studies using a number of methods of assessing the chemical control of breathing have shown a remarkable identity between these two twin athletes which allows the conclusion that, when the effect of a major environmental factor, physical activity, is removed, genetic factors predominate in determining the maximal oxygen uptake and the ventilatory responses to CO_2 and hypoxia.

Acknowledgments

I thank Professor K. W. Donald for laboratory facilities, Dr. Clarke of Glasgow for assistance, Miss E. Paxton for technical help and Miss L. Graham for typing the manuscript. I am grateful to Mr. C. Darg of the South Eastern Regional Blood Transfusion Service for the blood-grouping studies.

References

ARKINSTALL, W. W., NIRMEL, K., KLISSOURAS, V. and MILIC-EMILI, J. (1974) Genetic differences in the ventilatory response to inhaled CO_2. *Journal of Applied Physiology*, **36**, 6–11.

ASMUSSEN, E. and NIELSEN, M. (1947) Studies on the regulation of respiration in heavy work. *Acta Physiologica Scandinavica*, **12**, 171–188.

ASTRAND, P.-O. and RODAHL, K. (1970) *Textbook of Work Physiology*. New York: McGraw-Hill.

BYRNE-QUINN, E., WEIL, J. V., SODAL, I. E., FILLEY, G. F. and GROVER, R. F. (1971) Ventilatory control in the athlete. *Journal of Applied Physiology*, **30**, 91–98.

CUNNINGHAM, D. J. C., SHAW, D. G., LAHIRI, S. and LLOYD, B. B. (1961) The effect of maintained ammonium chloride acidosis on the relation between pulmonary ventilation and alveolar oxygen and carbon dioxide in man. *Quarterly Journal of Experimental Physiology*, **46**, 323–334.

FLENLEY, D. C., COOKE, N. J., KING, A. J., LEITCH, A. G. and BRASH, H. M. (1973) The hypoxid drive to breathing during exercise in normal man and in hypoxic patients with chronic bronchitis and emphysema. *Bulletin de Physiopathologie Respiratoire*, **9**, 669–671.

HIRSCHMAN, C. A., McCULLOUGH, R. E. and WEIL, J. V. (1975) Normal values for hypoxic and hypercapnic ventilatory drives in man. *Journal of Applied Physiology*, **38**, 1095–1098.

KLISSOURAS, V., PIRNAY, F. and PETIT, J.-M. (1973) Adaptation to maximal effort: genetics and age. *Journal of Applied Physiology*, **35**, 288–293.

KRONENBERG, R., HAMILTON, F. N., GABEL, R., HICKEY, R., READ, D. J. C. and SEVERINGHAUS, J. W. (1972) Comparison of three methods for quantitating respiratory response to hypoxia in man. *Respiration Physiology*, **16**, 109.

MITCHELL, J. H. and BLOMQUIST, G. (1971) Maximal oxygen uptake. *New England Journal of Medicine*, **284**, 1018–1022.

NICHOLS, R. C. (1965) The national merit twin study. In *Methods and Goals in Human Behaviour Genetics*, pp. 231–243. Ed. Vandenberg, S. G. New York: Academic Press.

READ, D. J. C. (1967) A clinical method for assessing the ventilatory response to carbon dioxide. *Australian Annals of Medicine*, **16**, 20–32.

REBUCK, A. S. and READ, D. J. C. (1971) Patterns of ventilatory response to CO_2 during recovery from severe asthma. *Clinical Science*, **41**, 13–31.

SALTIN, B. and ASTRAND, P.-O. (1967) Maximal oxygen uptake in athletes. *Journal of Applied Physiology*, **23**, 353–358.

SALTIN, B., BLOMQUIST, L., MITCHELL, J. H., JOHNSON, R. L., WIDDENTHAL, K. and CHAPMAN, C. G. (1968) Response to exercise after bed rest and after training. *Circulation*, **37** (Suppl. 7).

TAYLOR, H. L., BUSKIRK, E. and HENSCHEL, A. (1955) Maximal oxygen uptake as an objective measure of cardio-respiratory performance. *Journal of Applied Physiology*, **8**, 73–80.

WEIL, J. V., BYRNE-QUINN, E., SODAL, I. E., FRIESEN, W. O., UNDERHILL, B., FILLEY, C. G. and GROVER, R. F. (1970) Hypoxic ventilatory drive in normal man. *Journal of Clinical Investigation*, **49**, 1061–1072.

WEIL, J. V., BYRNE-QUINN, E., SODAL, I. E., KLINE, J. S., McCULLOUGH, R. E. and FILLEY, G. F. (1972) Augmentation of chemosensitivity during mild exercise in normal man. *Journal of Applied Physiology*, **33**, 813–819.

DETERMINANTS OF RESPIRATORY FUNCTION IN BOY AND GIRL TWINS

J. E. COTES, C. HEYWOOD and K. M. LAURENCE

MRC Pneumoconiosis Unit, Llandough Hospital, Penarth, Glamorgan
and
Department of Paediatric Pathology, University Hospital of Wales,
Heath Park, Cardiff

THE functional dimensions of the lung, including the ventilatory capacity, the lung volume, the gas transfer factor and its subdivisions, are positively correlated with stature and other indices of body size. In adults the function declines with age from about 25 years; before this, after allowing for stature, age does not contribute significantly to the overall variability. For a given stature the vital capacity is less in females than in males, and compared with people of European descent is less in most other ethnic groups. The evidence is reviewed elsewhere (Cotes, 1975). It demonstrates unequivocally that lung function is largely genetically determined. In addition, for ventilatory capacity Higgins and Keller (1975) have shown a family resemblance between fathers and sons, and between mothers and daughters; however, little is known of the mechanisms, including the possible role of factors which are correlated with the cardiac frequency response to exercise. This is of practical importance since the frequency response also reflects the state of training of the cardiorespiratory apparatus and, in some studies is negatively correlated with the functional dimensions of the lung (e.g. Cotes, 1976; Miller, Gilson and Saunders, 1976). The conclusion from the latter studies that a high level of habitual activity during adolescence has a beneficial effect upon the lung might be in error if there are other genetic effects which have not been taken into account. To illuminate this problem, measurements of lung function and the cardiac

77

frequency response to exercise have been made on boy and girl twins.

Subjects and Methods

The subjects were boy and girl twin pairs who were identified mainly through the birth records for the area but also through local contacts. The way in which they were approached, the total numbers and the reasons for some results not being complete or being excluded from the analysis, and also a description of the relationship of the lung function to stature for this group of children, are given elsewhere (Cotes *et al.*, 1973). The numbers included in the present study are given in Table 1.

TABLE 1.   Details of the subjects: Mean values and ranges.

	Number	Age (yr)	Height (m)	Weight (kg)	Fat-free mass FFM
Boys					
Identical twins	48	11·6	1·46	37·8	31·5
		(8–16)	(1·21–1·80)	(22·5–60·0)	(20·3–49·8)
Non-identical twins	48	11·46	1·46	37·4	30·6
		(8–15)	(1·22–1·75)	(21·5–75·5)	(19·1–53·5)
From boy/girl pairs	32	10·7	1·42	32·9	28·1
		(8–16)	(1·19–1·79)	(22·5–63·5)	(19·6–53·4)
Girls					
Identical twins	56	11·4	1·44	37·0	28·2
		(8–16)	(1·18–1·62)	(19·0–56·5)	(15·8–48·2)
Non-identical twins	48	11·0	1·44	35·6	27·4
		(8–15)	(1·15–1·67)	(19·5–65·0)	(16·9–47·4)
From girl/boy pairs	32	10·7	1·41	32·9	25·2
		(8–16)	(1·23–1·69)	(20·5–61·0)	(16·9–47·4)

The indices which were obtained on the subjects together with the abbreviations by which they may be identified in the other tables are listed in Table 2. Stature was measured using a Holtan digital anthropometer with the subject at full stretch, heels on the ground and the head in the Frankfurt plane. Measurements on the boys were made by an experienced observer (GRJ); those on the girls by one who was only recently trained. Measurements of skinfold thickness were measured using Harpenden skinfold calipers; these

TABLE 2. Mean values and standard deviations for the lung function of 44 boy and 34 girl twin pairs whose data were analysed in terms of principal components.

	Boys		Girls	
	Mean	SD	Mean	SD
Forced expiratory volume (FEV$_{1.0}$, l)	2·50	0·75	2·27	0·52
Forced vital capacity (FVC, l)	3·04	0·89	2·61	0·57
Peak expiratory flow rate (PEFR, l min^{-1})	338·9	81·05	348·7	8·63
Total lung capacity (TLC, l)	3·80	1·14	3·33	0·78
Inspiratory capacity (IC, l)	2·03	0·58	1·74	0·41
Expiratory reserve volume (ERV, l)	1·01	0·43	0·89	0·27
Functional residual capacity (FRC, l)	1·79	0·66	1·61	0·45
Residual volume (RV, l)	0·76	0·30	0·70	0·24
Transfer factor (Tl, mmol min^{-1} kPa^{-1})	7·24	2·16	6·39	1·57
Transfer factor ÷ lung vol. (K_{CO}, mmol min^{-1} kPa^{-1})	2·03	0·283	2·01	0·209
Diffusing capacity of alv. membrane (Dm, mmol min^{-1} kPa^{-1})	12·02	3·43	10·46	3·14
Volume of blood in alveolar capillaries (Vc, ml)	59·3	25·32	52·6	15·63

were applied at four sites, biceps, triceps, subscapular and supra-iliac, according to the procedure recommended for the International Biological Programme (Weiner and Lourie, 1969). From these measurements and the body weight, the percentage of body weight which is fat and hence the fat-free mass were estimated using the method of Durnin and Rahaman (1967).

The ventilatory capacity was measured as forced expiratory volume and forced vital capacity using a McDermott dry bellows spirometer and as peak expiratory flow rate using a Wright peak flow meter. The lung volumes and transfer factor were measured respectively by closed circuit spirometry with helium as the indicator gas and the single breath carbon monoxide method using a resparameter. The transfer factor was partitioned into its components the diffusing capacity of the alveolar membrane and the volume of blood in the lung capillaries by making measurements of transfer factor at two levels of alveolar oxygen tension and assuming a value for the reaction rate of carbon monoxide with oxyhaemoglobin. Full details of all these methods are given elsewhere (Cotes, 1975). The physiological response to exercise was obtained using a Lanooy cycle ergometer. The rate of cycling was 50 per min and the load on the pedals was increased by 10 watts each minute. The measurements were made each minute of ventilation volume, oxygen uptake and cardiac frequency from which were derived the ventilation and cardiac frequency at the oxygen uptake of 45 mmol min^{-1} (1·01 min^{-1}). Full details of the apparatus, procedure and calculations are given elsewhere (Cotes, 1972).

The results were analysed in two ways; first by partitioning the total variance into components attributable to the genetic and non-genetic (environmental) components (Falconer, 1960), and second, by condensing the lung function results through the procedure of principal component analysis; the main components were analysed with respect to height, fat-free mass, exercise cardiac frequency and sex, also to the magnitudes of the differences within twin pairs.

Results

Details of the subjects are given in Table 1. They comprise 264 children all of whose results were used to derive the components of the principal component analysis. The analysis of the heritability of stature was applied to the data for all the single-sex twin pairs. That of the lung function expressed in terms of principal components was confined to those pairs where both children attained an oxygen uptake of or near to 45 mmol min^{-1}, such that the cardiac frequency at this rate of energy expenditure could be ascertained. A total of 44 boys and 34 girl pairs met this requirement; the mean values of the indices of lung function for these children are given in Table 2.

Estimates of the genetic correlation, made on the assumption that the environmental correlation is the same for both monozygous and dizygous pairs, are given in Table 3. This shows that the precision of the estimate depends critically on the method adopted for obtaining the variance with respect to age. A curvilinear relationship of height on age to the nearest day, sub-divided by age-group gives the highest correlation in boys of 0·84 and in girls of about 0·5. These results may be compared with the generally accepted value of 0·9 estimated for example by Huntley (1966).

The results of the principal component analysis are given in Table 4, which contains the constituent of the three components arranged in descending orders of the magnitude of the contribution of each index of lung function. The first component reflects the size of the lung, the second its capacity to transfer gas whilst the third

TABLE 3. Heritability of stature in boy and girl twins.

	Boys	Girls
Ignoring age	0·19	0·17
Age adjusted		
linear	0·68	0·49
within age groups	0·76	0·54
cubic (age to the day)	0·84	0·48

The generally accepted value is 0·9 (Huntley, 1966).

contains information about the gas-exchanging properties of the lung. Some attributes of the components are given in Table 5. Component 1, which accounts for 73 per cent of the total variability within the data, is strongly correlated with indices of body size including stature and fat-free mass. The second component is independent of body size and this is partly the case for the third component. Both these components are positively correlated with the exercise cardiac frequency, whilst all the components are sex-linked being larger in boys than girls. The magnitude of the within pair differences in the components is given in Table 6 where the components have been standardized to the mean height, fat-free mass and cardiac frequency for children of each sex separately. The table shows that the within pair differences for all three components are

TABLE 4. Principal component analysis*.

1st		2nd		3rd	
TLC	1·00	K_{CO}	1·00	Vc	1·00
FEV	0·98	Dm	0·45	Dm	−0·82
FVC	0·97	Tl	0·39	RV	0·41
FRC	0·94	RV	0·20	PFR	−0·28
Tl	0·93	ERV	−0·18	IC	−0·20
IC	0·89	FEV	−0·13	FRC	0·17
ERV	0·86	FVC	−0·11	K_{CO}	0·14
PFR	0·84	PFR	−0·11	FEV	−0·14
RV	0·81	Vc	−0·07	FVC	−0·13
Vc	0·80	IC	−0·06	ERV	0·06
Dm	0·78	TLC	−0·04	TLC	0·02
K_{CO}	−0·33	FRC	−0·02	Tl	0·01

* except for PFR the constituents are in logarithmic form.

Table 5.   Properties of components.

	1	2	3
% of total variability	73·4	10·2	4·8
Relation to Ht	+	nil	+
Relation to FFM	+	nil	nil
Relation to $fC_{1·0}$	nil	+	+
Relation to $V_{E1·0}$	nil	nil	+
Relation to sex	♂ > ♀	♂ > ♀	♂ > ♀

similar as between the mono- and dizygous twin pairs. By contrast, the exercise cardiac frequency standardized for fat-free mass exhibits significantly greater differences as between dizygous than between monozygous pairs.

Discussion

The present data have proved difficult to interpret on account of the high correlation between lung function and stature so that the range of ages of the children comprising the sample and hence the range of statures, critically determines the overall group variance. In these circumstances, the fact that the heritability of stature, which may be derived from the data in the case of the boys, only just attains the accepted level, whilst the estimated heritability of the girls is well below this level, militates against being able to extract a reliable estimate of heritability for lung function independent of stature.

Table 6.   Differences within twin pairs in principal components 1, 2 and 3 standardized for ht, FFM and $fC_{1·0}$, also $fC_{1·0}$ standardized for FFM.

		Boys		Girls	
		Monozygous	Dizygous	Monozygous	Dizygous
$\Delta PC1$, st		0·439	0·679	0·613	0·796
	SD	0·392	0·437	0·411	0·395
$\Delta PC2$, st		0·850	0·799	0·796	1·063
	SD	0·484	0·631	0·467	0·824
$\Delta PC2$		0·786	0·779	0·844	0·128
	SD	0·463	0·648	0·466	0·828
$\Delta PC3$, st		0·560	0·892	0·599	0·525
	SD	0·496	0·840	0·328	0·467
$\Delta fC_{1·0}$*		7·5	13·6	0·6	11·3
	SD	4·29	8·97	0·44	10·1

* $P < 0.05$

However, the lung function results are of high technical quality for both sexes and the lower heritability for the girls is possibly due to the estimates of stature being less precise than is the case for the boys, so the results justify further analysis.

The use of principal component analysis was intended to eliminate the effect of stature by concentrating its variance in the first component and this turned out to be the case. For none of the first three components were the within-group differences larger for the dizygous than for the monozygous pairs either before or after standardization to common values within pairs for body size and exercise cardiac frequency. Since these components account for 88 per cent of the total variability in the lung function results the similarity within pairs suggests that after allowing for stature the remaining variability is mainly of non-genetic origin. Its precise nature is unclear but the absence of a significant negative correlation with cardiac frequency suggests that habitual activity is not an important contributor. This result differs both from those of Klissouras (1977) who finds a high h index for vital capacity divided by height and with the expectation from human biological studies, including those cited in the introduction which show an association between lung function and exercise cardiac frequency. However, whereas in these studies the subjects were drawn from different environments, for example, rural and urban children in Jamaica (Miller et al., 1976) the present subjects were relatively homogeneous with respect to habitual activity. They may also have been more similar in other respects since the environmental differences which were an integral part of the other studies may have lead to selective mortality and/or migration and so to genetic differences as well. Comparison of the two sets of results suggests that the level of activity is either an unimportant determinant of lung function or needs to be material if it is to exert an effect.

For exercise cardiac frequency, the larger within pair differences for the dizygous compared with the monozygous twins suggests that this index, which reflects the maximal oxygen uptake (Cotes et al., 1969), may be mainly genetically determined. This is also the conclusion reached by Klissouras, Pirnay and Petit (1973) but not by Howald (1977). However, our experience with the lung function data suggests that the difference between monozygous and dizygous pairs may equally be due to inappropriate allowance having been made for body size or to the pairs differing with respect to habitual activity. The physiological evidence for environmental factors affecting this and other indices of physical fitness is so strong that the genetic explanation must still be considered speculative.

Conclusions

Our findings support the generally accepted view that body size is the main determinant of lung function in children. Since stature is mainly genetically determined this must also be the case for lung function. After allowing for size the lung function is different between the sexes and between ethnic groups. It has not been possible to demonstrate a genetic component in addition to those which relate to size, sex and ethnic group. Exercise cardiac frequency is more similar with monozygous than within dizygous twin pairs. Evidence as to the cause is incomplete.

Acknowledgments

We are indebted to Dr. P. D. Oldham for the analysis of heritability of stature, to Mr. G. Berry for advice on the principal components analysis and to our many colleagues whose contributions to the collection of the data made this analysis possible.

References

COTES, J. E. (1972) Response to progressive exercise: a three index test. *Br. J. Dis. Chest.*, **66**, 169–184.

COTES, J. E. (1975) *Lung Function. Assessment and Application in Medicine*, 3rd edition. Blackwell Scientific Publications, Oxford.

COTES, J. F. (1976) Genetic and environmental determinants of the physiological response to exercise. *Medicine and Sport*, **9**, 93–107.

COTES, J. E., DABBS, J. M., HALL, A. M., AXFORD, A. T. and LAURENCE, K. M. (1973) Lung volumes, ventilatory capacity and transfer factor in healthy British boy and girl twins. *Thorax*, **28**, 709–715.

COTES, J. E., DAVIES, C. T. M., EDHOLM, O. G., HEALY, M. J. R. and TANNER, J. M. (1969) Factors relating to the aerobic capacity of 46 healthy British males and females, ages, 18 to 28 years. *Proc. R. Soc. Lond.* B, **174**, 91–114.

DURNIN, J. V. G. A. and RAHAMAN, M. M. (1967) The assessment of the amount of fat in the human body from measurements of skinfold thickness. *Br. J. Nutr.*, **21**, 681–689.

FALCONER, D. S. (1969) Introduction to cumulative genetics. Oliver and Boyd, Edinburgh.

HIGGINS, M. and KELLER, J. (1975) Familial occurrence of chronic respiratory disease and familial resemblance in ventilatory capacity. *J. chron. Dis.*, **28**, (4) 239–251.

HOWALD, H. (1977) Ultrastructure and biochemical function of skeletal muscle in twins. This volume, pp. 57–66.

Huntley, R. M. C. (1966) A study of 300 twin pairs and their families showing resemblances in respect of a number of physical and psychological measurements. Ph.D. Thesis, University of London, Institute of Child Health.

Klissouras, V. (1977) Twin studies on functional capacity. This volume, pp. 43–55.

Klissouras, V., Pirnay, F. and Petit, J. M. (1973). Adaptation to maximal effort: genetics and age. *J. appl. Physiol.*, **35,** 288–293.

Miller, G. J., Saunders, M. J., Gilson, R. J. and Ashcroft, M. T. (1976) Superior lung function and exercise performance of Jamaican children belonging to a hill farming community as compared with their urban compatriots. *Thorax*, **31,** 481–482.

Weiner, J. S. and Lourie, J. A. (1969). *Human Biology: a Guide to Field Methods.* Blackwell Scientific Publications, Oxford.

THE INTERACTION OF GENETIC AND ENVIRONMENTAL FACTORS IN DETERMINING THE RESEMBLANCE OF ARTERIAL PRESSURE IN CLOSE RELATIVES

W. E. MIALL

MRC/DHSS Epidemiology and Medical Care Unit, Northwick Park Hospital

ARTERIAL pressure is one of the most variable of human physiological measurements, as can be recognized by an examination of any of the 24 hour records that are produced by the new continuous recorders. There is a general tendency for pressure to increase during the day—mean evening values being a few mm Hg higher than mean morning values. Systolic pressure varies than diastolic pressure, and in many people the highest systolic pressure during the 24 hours exceeds twice the lowest value; in some hypertensive subjects systolic pressure by night may be lower than diastolic pressure by day.

The variability of blood pressure tends to increase with its mean value—the higher the pressure the greater the lability both absolutely and proportionately. The concept that, as a person's pressure rises, he passes from a normotensive stage through a phase of labile hypertension to fixed hypertension is a myth, resulting from the use of some arbitrary threshold to define abnormal from normal pressure. Nevertheless, the response to blood pressure raising stimuli itself shows biological variation and there is little doubt that this is partly determined by familial factors. The defence reflex, emotion, pain, coitus and a full bladder are some of the short-term stimuli which raise arterial pressure; posture usually has relatively little effect though there is a small drop in systolic pressure and a

87

small rise in diastolic pressure on taking up the recumbent position. Exercise is usually accompanied by a rise in both systolic and diastolic pressure, and by a greater lability of pressure.

With all these short-term influences it is not surprising that the inheritance factor has been found difficult to study. The earliest studies were based on family histories alone and, for a condition as common as essential hypertension, impressive family histories of its complications were frequently found but were of doubtful value. Soon after the invention of the sphygmomanometer at about the turn of the century, family studies based on measurements were undertaken but were designed according to the belief prevailing at the time that hypertension could be defined in some scientifically meaningful way in terms of a threshold separating normal from abnormal values; one either had the condition or one did not have it. A search through anatomical, physiological, epidemiological, pharmacological, clinical or pathological evidence fails to provide support for this concept.

Figure 1 shows standardized mortality ratios in middle-aged men, from a large insurance study, according to systolic and diastolic pressure (Society of Actuaries, 1959). Within any range of diastolic pressure, mortality increases with each increment of systolic pressure, and within any systolic pressure range it increases with diastolic pressure. There is no suggestion in such data of any critical values at which mortality suddenly increases.

The difficulty of introducing any arbitrary definition of hypertension can be illustrated by looking at the prevalence by age and sex in the general population. Figure 2 shows such data for 35–74 year old adults in the general population in South Wales, based on single casual diastolic (IV) measurements, and defining hypertension in three ways—90 mm Hg and more, 100 and more and 110 and more. The prevalence of hypertension can obviously be made to agree with any preconceived ideas with a little manipulation of definition, or of age and sex groups. Use the mean of multiple measurements instead of single ones and it can be made to vary enormously.

The results of the early inheritance studies, because of the arbitrariness of definitions, accorded with the current views of the nature of the inheritance and of the gene frequency (Weitz, 1923; Søbye, 1968). They were consistent with the hypothesis that essential hypertension

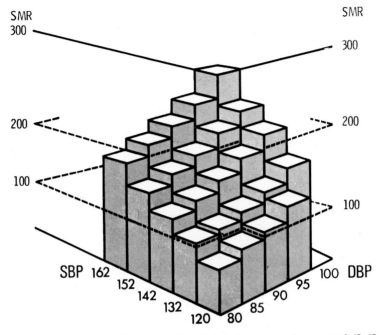

FIG. 1.   Mortality according to systolic and diastolic pressure, in men aged 40–69 (at issue).
Adapted from Society of Actuaries "Build and Blood Pressure" study, 1959.

was transmitted by a single pair of genes with features of Mendelian dominance. This implied that those who had inherited the gene differed from those that had not by having some specific biochemical fault, and much time and effort has since been devoted to a search for this biochemical characteristic, so far without success.

The concept of a single, specific, genetically determined explanation for essential hypertension was challenged by Hamilton, Pickering, Roberts and Sowrey (1954 a, b, c) who showed that no natural boundary separated abnormal from normal blood pressure and that the distributions of blood pressure were unimodal, both in the general population and in the close relatives of patients selected from the upper and the lower part of the blood pressure range. They postulated that inter-individual differences in blood pressure were of degree rather than of kind, and likely to be determined by polygenic inheritance and much influenced by environmental factors.

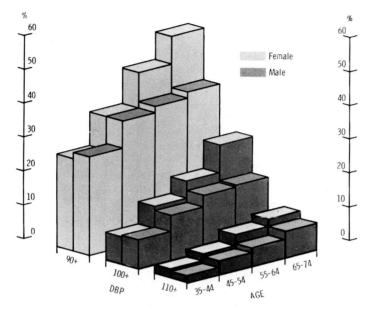

FIG. 2. The prevalence of diastolic (IV) hypertension of 90 mm Hg and more, 100 mm Hg and more, and 110 mm Hg and more, by age and sex. (Rhondda Fach and Vale of Glamorgan.)

Up to this time all the studies of the inheritance of blood pressure had started with propositi who were selected as either hypertensive or normotensive. Studies based on families unselected in terms of blood pressure levels were clearly needed, and a number of epidemiological studies of general populations started in the 1950's (Miall and Oldham, 1958; Cruz-Coke, 1959; Johnston, Epstein and Kjelsberg, 1965). In the absence of any agreement on how to define hypertension, techniques appropriate to the study of continuously distributed, quantitative characteristics (such as intelligence quotients) were used. These techniques are appropriate for the study of polygenically determined traits, and they were used on the assumption that the analyses of data would reveal evidence incompatible with polygenic inheritance, if that hypothesis were incorrect. If correct, however, the accumulation of data, and more reliable data, would produce analyses which should accord increasingly well with what would be expected from polygenic inheritance.

Though similar results have been described from other studies, for convenience, I will illustrate from our own data in South Wales. Random samples of the general population (aged 5 years and over) from two areas in South Wales were drawn and used as propositi for family studies which included all first degree relatives living within access. There were 623 families which were observed at intervals for between 15 and 17 years. The measurements were made by one observer in the individuals' homes, and were analysed in the appropriate way for investigating polygenic inheritance. Age and sex adjusted scores were derived for each subject; they represented the deviation of that person's pressure from the mean for his or her 5 year age and sex group, expressed in standard deviation units. There are weaknesses in such scores. Implicit in such a method of analysis is the assumption that the average men or woman follows, as his blood pressure rises with the passage of time, the curve relating pressure with age in his or her sex. This is manifestly not true. Also implicit in this scoring system is the hypothesis that only non-genetic influences can cause an individual to change his position *vis-à-vis* that of his peers in the hierarchy of blood pressures in his age group.

Using such scores, the relationship in blood pressure between propositi and relatives was examined by linear regression methods after the initial survey of the first population studied—that of the Rhondda Fach. Figure 3 shows the considerable, but not statistically

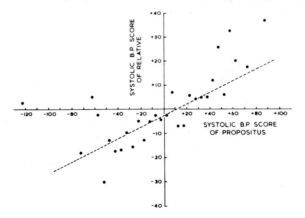

FIG. 3.   The relationship between the systolic blood pressure scores of propositi and the mean systolic scores of relatives.

significant, scatter about the regression line for systolic pressure. A similar analysis was subsequently undertaken using the data from the two populations, pooled, (Fig. 4) and clearly the fit with the regression had improved. When the means of the measurements

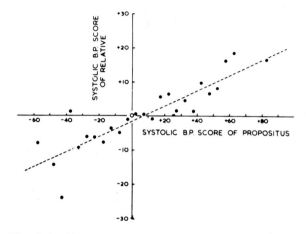

FIG. 4.   The relationship between the systolic blood pressure scores of propositi and the mean systolic scores of relatives.

made at the first and second surveys of these two populations were analysed, the fit improved again (Fig. 5). Data for diastolic pressure showed a similar picture. These findings accorded well with the concept of polygenic inheritance; as more data and more reliable

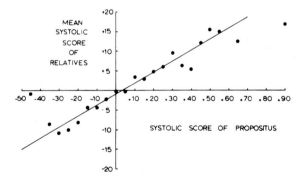

FIG. 5.   The relationship between the systolic blood pressure scores of propositi and the mean systolic scores of relatives.

data became available for analysis, they fitted better with the pattern expected on this hypothesis (Miall and Oldham, 1963). No evidence suggested that the pattern of inheritance differed between those with high, average or low pressures, and there was no convincing evidence of bimodal distributions in the scores of their relatives (Fig. 6).

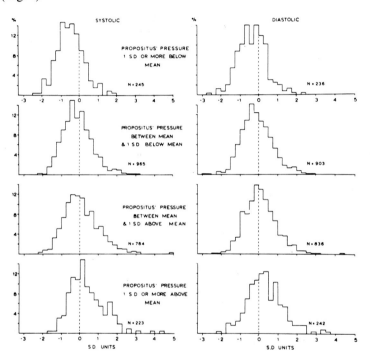

FIG. 6.   Distributions of arterial pressure (standard deviation unit) scores of relatives according to those of propositi.

The degree of resemblance between propositi and relatives in blood pressure can be represented by a regression of about 0·29 for systolic pressure, and 0·22 for diastolic pressure. If translated into more familiar terms, this means that the 1st degree relatives of a person with a score of + 50 mm Hg, i.e. 50 mm Hg above the average for age and sex, will on average have pressures 14·5 mm (systolic) and 11 mm (diastolic) above the mean at the equivalent age.

To examine the relevance of these epidemiological findings to the

problem as seen by those in hospital dealing with severely hypertensive patients, the data of Platt (1963) and Pickering were scored,
using the scores derived from the general population in South
Wales. For both systolic and diastolic pressure the data from
severely hypertensive propositi fitted those of the general population up to scores of + 60, which included about 60 per cent of the
clinic families for systolic and 75 per cent for diastolic pressures (Figs.
7 and 8). At higher levels, the mean scores of relatives lag behind
those of their propositi suggesting, perhaps, that those with severe
hypertension resemble their relatives less than they did at an earlier
stage.

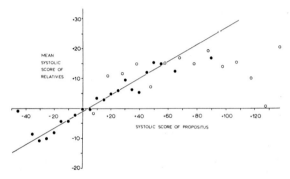

FIG. 7.   The relationship between the mean systolic scores of relatives and those of
propositi.
● General population, Wales.
○ Data from cases of essential hypertension (Hamilton *et al.*, 1954 ; Platt, 1963).

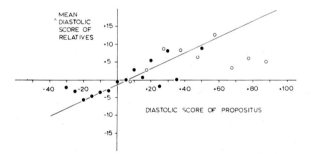

FIG. 8.   The relationship between the mean diastolic scores of relatives and those of
propositi.
● General population, Wales.
○ Data from cases of essential hypertension (Hamilton *et al.*, 1954 ; Platt, 1963).

In summary, then, the epidemiological data appear to fit the polygenic hypothesis reasonably well, though population studies cannot exclude the possibility that a small proportion of hypertension may be transmitted by single gene inheritance as in the case of stature. If it is, then it is too small to be detected using our present techniques and certainly cannot account for the major problem of hypertension and its clinical complications experienced in the general population.

The extent to which this familial resemblance in blood pressure is genetic rather than a reflection of shared environments has been explored by twin studies and studies of spouses. Twin studies show a consistently greater correlation between the pressures of monozygotic than dizygotic twins (Stocks, 1930) some spouse studies have (Chazan and Winkelstein, 1964), but other have not (Miall *et al.*, 1967), shown resemblance in pressure even in those who have shared the same home environment for at least 10 years.

If blood pressure were determined by polygenic inheritance all types of first degree relatives would be expected to resemble each other to the same extent. In our own studies, many more regression coefficients (relating pressures of different types of kin with propositi) than could be expected by chance were found to differ significantly (Miall *et al.*, 1967) (Table 1). There were features in the pattern of this heterogeneity which suggested that some genes controlling blood pressure were situated on the x-chromosome and the possibility of sex linkage was explored using the techniques of Finney (1939), but with equivocal results. It seemed more probable that the significant heterogeneity in the regression coefficients between different classes of relationship resulted from interaction between genetic and environmental influences, and that these environmental factors were different in the two sexes as judged by the significant changes in these regressions with age (Table 2).

Clustering of blood pressure levels has been demonstrated by Zinner, Levy and Kass (1971) between mothers and children aged less than two years. Some evidence suggests that it is the level of blood pressure which determines its own rate of increase (Miall and Lovell, 1967; Evans and Rose, 1971), probably by initiating and maintaining hypertrophy of the media of arteries; it is therefore possible that those who are destined to become hypertensive as adults could be detected in infancy or early childhood. A search for

TABLE 1.   Regression coefficients of relatives S.D. scores for systolic and diastolic pressure on those of male and female propositi. Rhondda Fach and Vale of Glamorgan

Relations	Male propositi			Female propositi		
	Regression coefficients			Regression coefficients		
	No.	Systolic	Diastolic	No.	Systolic	Diastolic
Father	139	0·164	0·118	134	0·328	0·164
Mother	166	0·434	0·418	153	0·171	0·149
Brother	299	0·296	0·204	246	0·285	0·306
Sister	264	0·303	0·166	279	0·330	0·326
Son	132	0·139	0·091	140	0·203	−0·014
Daughter	137	0·216	0·235	269	0·413	0·371
All Male relatives	570	0·234	0·158	509	0·274	0·188
All Female relatives	567	0·324	0·257	572	0·319	0·299
All relatives	1137	0·277	0·208	1081	0·298	0·244

All relatives on all propositi: systolic 0·286 ± 0·022; diastolic 0·225 ± 0·022.

factors causing familial aggregation of blood pressure might be expected to be more rewarding in those showing early manifestations of it, perhaps less influenced by environmental interaction, and investigations of the nature of the inheritance factor are now being directed more towards infants and young children. With hindsight,

TABLE 2.   Regression coefficients for systolic and diastolic S.D. scores of relatives on those of propositi of different ages. Rhondda Fach and Vale of Glamorgan (1967).

	Propositi under age 15		Propositi aged 15–44		Propositi aged 45 and over	
	No.	Regression	No.	Regression	No.	Regression
Systolic						
Relatives of same sex as propositus	224	0·213	561	0·282	357	0·306
Relatives of opposite sex to propositus	220	0·402	519	0·304	337	0·234
Diastolic						
Relatives of same sex as propositus	224	0·154	561	0·220	357	0·283
Relatives of opposite sex to propositus	220	0·396	519	0·208	337	0·149

it is perhaps not surprising that in a complex situation where environmental factors interact—as they always do—with genetic ones, where blood pressure levels are influenced by morbidity and by therapy, where there are sex differences in survival of hypertension and where the characteristic is one showing inate variability, population studies have been unable to provide a very accurate measure of the extent to which familial aggregation is genetic. Present evidence, however, suggests that about 55–60 per cent of the variance of systolic pressure, and 45 per cent of that of diastolic pressure, is accounted for by familial factors. Non-familial environmental factors, some of which will contribute to the minute-by-minute variations of pressure, presumably account for the remainder.

References

CHAZAN, J. A. and WINKELSTEIN, W. (1964) *J. chron. Dis.*, **17**, 9.

CRUZ-COKE, R. (1959) *Acta genet. Statist. med.*, **9**, 207.

EVANS, J. G. and ROSE, G. (1971) *Br. med. Bull.*, **27**, 37.

FINNEY, D. J. (1939) *Ann. Eugen.*, **9**, 203.

HAMILTON, M., PICKERING, G. W., ROBERTS, J. A. F. and SOWREY, G. S. C. (1954 a) *Clin. Sci.*, **13**, 11.

HAMILTON, M., PICKERING, G. W., ROBERTS, J. A. F. and SOWREY, G. S. C. (1954 b) *Clin. Sci.*, **13**, 37.

HAMILTON, M., PICKERING, G. W., ROBERTS, J. A. F. and SOWREY, G. S. C. (1954 c) *Clin. Sci.*, **13**, 273.

JOHNSON, B. C., EPSTEIN, F. H. and KJELSBERG, M. O. (1965) *J. chron. Dis.*, **18**, 147.

MIALL, W. E., HENEAGE, P., KHOSLA, T., LOVELL, H. G. and MOORE, F. (1967) *Clin. Sci.*, **33**, 271.

MIALL, W. E. and LOVELL, H. G. (1967) *Br. med. J.*, **1**, 660.

MIALL, W. E. and OLDHAM, P. D. (1958) *Clin. Sci.*, **17**, 409.

MIALL, W. E. and OLDHAM, P. D. (1963) *Br. med. J.*, **1**, 35.

PLATT, R. (1963) *Lancet*, **1**, 899.

SOCIETY OF ACTUARIES (1959) *Build and Blood Pressure Study*, Vol. 1. Chicago: Society of Actuaries.

SØBYE, P. (1948) *Op. Domo Biol. Hered. Num. Kbh.* Vol. 16.

STOCKS, P. (1930) *Ann. Eugen.*, **4**, 49.

WEITZ, W. (1923) *Z. Klin. Med.*, **96**, 151.

ZINNER, S. H., LEVY, P. S. and KASS, E. H. (1971) *New Engl. J. Med.*, **284**, 401.

HANDEDNESS AND THE CEREBRAL REPRESENTATION OF SPEECH

MARIAN ANNETT

Department of Applied Social Studies, Lanchester Polytechnic, Coventry

Introduction

PRECISE hand movement and speech are two of man's most highly skilled performances. In most people they appear to go together in the sense that the preferred right hand and speech are controlled by the left hemisphere. Whether this conjunction depends on coincidence or on some intrinsic connection is unknown. It is clear that a simple contralateral relationship between speech hemisphere and handedness does not hold as there are many left-handers with left cerebral speech and some right cerebral speakers who are right-handed. But handedness and brainedness are not independent. The incidence of left-handedness is much higher among the right-brained than among the left-brained.

The relation between the lateral asymmetries of hand and brain is a problem at the centre of many associated problems about other human and non-human lateral asymmetries of structure and function. It is not possible to investigate this core problem directly at present. since there is no safe, convenient and reliable method of assessing cerebral speech in the general population. The main evidence depends on patients having unilateral right or left cerebral insults, who are assessed for handedness and for dysphasia. Small samples of such patients can be used to test limited hypotheses. It is easy to disprove, for example, the proposition that all left-handers are right-brained for speech, or to demonstrate that patients with right cerebral speech have a high proportion of sinistral relatives. But if the relations between handedness and brainedness are to be studied as human species characteristics, it is essential to have samples unselected for handedness and brainedness and large enough to have several instances of the more infrequent combinations. These criteria are not easily fulfilled. Five large consecutive series have been identified in the literature by Zangwill (1967). These series have been re-analysed (Annett, 1975) in order to test predictions about the relations between handedness and brainedness which arose from a theory of the origins of handedness (Annett, 1972). It is the purpose of this paper to outline the theory and to summarize its application to the data on dysphasia and handedness.

Hand Preference

The theory about the determinants of handedness developed through several stages, of which the first few are represented schematically in Fig. 1. Starting at the top and proceeding downward as in an archaeological dig, successive layers of the problem are uncovered. The structure of the problem becomes simpler as one descends. The difficulty for exposition is that if one started with the bare bones, they would have no obvious resemblance to the familiar surface features. The starting point is the dichotomous classification into right- and left-handers used in most of the literature, including that concerning dysphasia. It will be necessary to return to this level to examine the evidence after analysing the structure of the problem below the surface.

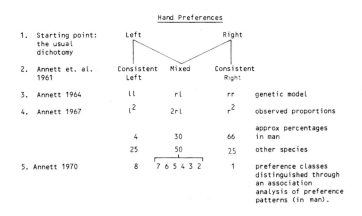

FIG. 1.   Stages in the analysis of hand preferences.

The findings of Annett, Lee and Ounsted (1961) on the abilities of epileptic children with unilateral EEG foci seemed to demand a classification of patients into three groups, distinguishing consistent handers right and left from mixed handers, who prefer the right hand for some skilled unimanual actions and the left hand for others. In subsequent studies of schoolchildren and undergraduates, mixed-handedness was found to be common (in about 30 per cent) while true ambidexterity in the sense of use of either hand for a skilled activity like writing was rare (in about 0·34 per cent).

Annett (1964) suggested that the three varieties of handedness, right, mixed and left, might depend on a simple two-allele genetic model in which left-handedness depended on a recessive gene with partial penetrance in heterozygotes. If some heterozygotes were variable in hand preference and also variable in cerebral speech laterality, but arbitrarily classified as right- or left-handed, exceptions to the contralateral relation between handedness and brainedness might be accounted for. Rife (1950) had suggested that hetero-zygosity might underlie the discordance of right- and left-handed monozygotic twins. As Rife, and most others in the field, did not recognize the variability of preference shown by mixed handers, it was necessary to re-examine questions about the distribution, development, genetics and neurology of handedness. Research was begun on several of these problems.

When samples of children and undergraduates, unselected for handedness, were observed performing several actions thought to be sensitive to lateral preferences, or were questioned about these actions, consistent right, mixed and consistent left handers were in proportions compatible with the random assortment of two alleles and full expression of both genes in heterozygotes. At the same time, it was evident from family data in which parents and children were classified as required to test the model that the simple two-allele model was wrong; two consistently right-handed parents in whom no traces of mixed handedness could be discerned, could have left-handed children. Thus, Annett (1967) described the findings of binomial proportions but could not explain them. Binomial pro-portions were found in several samples in the literature when suffi-cient data was given for re-analysis. Even more surprisingly, they were found in the distributions of paw preference reported for mice, rats, cats, monkeys and chimpanzees. Although the percentages in these species are very different from those found in man, the non-human approximation to 25, 50, 25 per cent left, mixed and right is in binomial proportion as much as the human 4, 30, 66 per cent.

The 12-item questionnaire used routinely in collecting information about hand preferences in undergraduates and their parents was subjected to an association analysis to see if any objective criteria could be discovered for distinguishing between preference patterns. Preferences were found to vary in an enormous number of patterns with all possible degrees of mixed usage (Annett, 1970). It was concluded that the preference distribution is continuous, not discrete. It was possible to delineate several sub groups of mixed handers, roughly ordered on the preference continuum, but where the continuum is divided is arbitrary.

Manual Skill

Simultaneously with the collection of data on preference, measurements were made of the skill of each hand with the peg-moving task shown in Fig. 2. The time taken to move the pegs from the top row to the bottom row was measured by stop-watch and the mean calculated for from 3 to 5 trials with each hand. The distribution of differences between the hands in movement time appeared to resemble that reported for the data collected by Galton on the strength of each hand (Woo and Pearson, 1927). The distribution was continuous, unimodal, roughly normal and with a mean to the right of the point of symmetry or zero difference between hands. Data on preference and skill in the same subjects made it clear that these two variables are closely related. Figure 3 shows the mean difference between hands for subjects in the original Hull sample divided into the 8 preference classes distinguished through the association analysis. It also shows new data for over 800 students attending Open University summer schools. The measurements were made by students testing one another during laboratory classes, in both samples. These data are discussed in detail elsewhere (Annett, 1976). The importance point is that for most preference classes, the mean differences between hands are virtually identical in the two samples. Handedness, which has seemed a slippery phenomenon for scientific enquiry, can now be regarded as anchored to differences between the hands in skill. Although there is variability between individuals within preference groups, it is possible to consider the general relationships between preference and skill in the total population.

FIG. 2.   The peg-moving task.

Figure 4 shows an idealized representation of the probable relationship between preference and skill, ignoring the overlaps between groups which would be found in any empirical study (Annett, 1972). Within a certain range of the point of zero difference between hands in skill, we would expect individuals to show mixed hand preferences. Beyond these thresholds, on either side of zero, we would expect consistent preferences, right and left. The 4, 30 and 66 per cent found in preference groups in human samples are consistent with expectations for these areas under the normal curve. Having looked at the human data in this way, it is easy to see that if laterality in other species also depended on a normal distribution of differences in skill, and on similar thresholds for mixed and consistent preference, but with a mean at 0, the 25, 50, 25 per cent preferences in animals would be accounted for. The proportions in all species seem to be a function of the range of liability to mixed handedness, and the fact that when a normal distribution is shifted a short distance over this range, the areas under the curve remain roughly

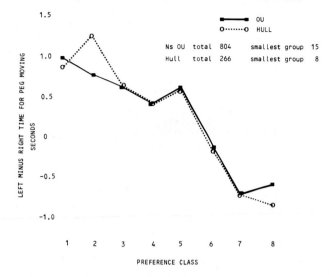

FIG. 3.   The relation between preference class and difference between hands in skill in two samples.

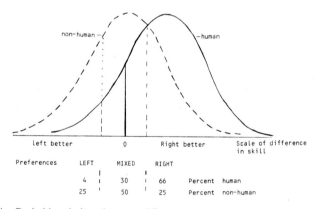

FIG. 4.   Probable relations between differences between hands in skill and hand preference.

in binomial proportion. The proportions, thus accounted for, do not matter in themselves but they point to the important conclusion that there is no significant difference between the human and non-human samples except that the human distribution is shifted in a dextral direction.

A Theory

There now seem to be only two things to be explained. The first is the origin of the distribution itself, which appears to be similar in all species known to manifest forelimb preferences. The second is the human shift to the right. Attempts to breed for lateral preferences in rats (Peterson, 1934) and in mice (Collins, 1968, 1969) have met with no success. In these species the distribution of differences probably arises from accidental and non-genetic causes. There must be many minor chance variations in the development of motor and sensory systems on the two sides of the body and there would be equal chances of greater skill on the right and the left, with the majority of animals showing no marked bias to either side; the animal data are fully consistent with this hypothesis. If this is the origin of the distribution of differences in other species, why not in man also? The hypothesis needs to be tested through human family studies which it has not been possible to persuade any research fund to support; the problem as now posed, seems too far from the surface to be immediately recognized as relevant. Some unpublished family data, collected with the help of my teenage children as research assistants, largely support the inference that the distribution of differences is of accidental origin.

The human shift to the right must depend on some systematic influence. This could be a universal human species characteristic or it could be subject to genetic variation; if genetic, it could be graded or discrete. There is no way of distinguishing between these possibilities at present. In what follows, it is assumed that the human shift to the right depends on a genetic factor which can be present or absent. When present, some advantage is given to the left hemisphere which is thereby induced to serve speech and which incidentally weights the accidental distribution of differences between the hands in favour of the right hand. This account assumes no intrinsic connection between handedness and cerebral speech. The dextral bias may be insufficient to outweigh very strong accidental biases to the left hand in some individuals who would then have left hemisphere speech and left hand preference.

The interesting predictions about handedness and brainedness arise from the possibility that in some individuals the factor giving a left hemisphere advantage is absent (RS— for absent and RS+ for present). Despite assiduous search, no evidence has been found for systematic tendencies to left hand preference or to right hemisphere speech in any group. When bias to the left hemisphere and right hand are absent, there is almost invariably a lack of bias to either side. In the children of two left handed parents the distribution of differences between the hands in skill was at least as wide as in unselected children but with a mean close to zero (Annett, 1974). This is consistent with the idea that the genetic bias to the right is absent in some families. Assuming, then, that there are some individuals who are RS—, what can be suggested about their handedness and cerebral speech? The simplest assumptions are that speech laterality and handedness depend on chance and on chances which are independent of each other.

Figure 5 summarizes the distributions now assumed by the theory. The parameters are unknown, but in this figure 80 per cent of the population are represented as RS+ with a mean difference between the hands in skill shifted $1\frac{1}{2}$ SD units to the right of 0. Twenty per cent are represented as RS—, their mean difference being, by definition, zero. Variances are the same in both subgroups. The total population is represented by the dotted line. It is roughly normal, with mean to the right of zero and a slight negative skew.

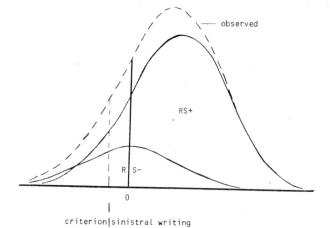

FIG. 5.   Hypothesized distributions.

When the samples of undergraduates and schoolchildren became larger, a significant negative skew was found in the distribution of differences between hands. Hoadley and Pearson (1929) compared the right and left internal lengths of some 700 skulls from ancient Egypt, and found a distribution of differences with similar characteristics.

There is a feature of Fig. 5 to which careful attention must be given. This is the threshold, or criterion of left-handedness in the total population. It is through the criterion that we can return to our starting point with the dichotomous classification of right- and left-handers. It has been shown that hand preferences vary continuously and are linked with a continuum of differences between the hands in skill. Any division of the continuum is arbitrary, but just where the division is made is a very significant variable in any empirical study. When several actions are enquired about, the criterion can range between left preference for *all* or for *any* actions, between about 3–35 per cent. Predictions about the proportions of right- and left-brainedness depend intimately on the criterion of left-handedness. These proportions have been estimated for several possible values of the criterion, given the parameters of Fig. 5 (Annett, 1975). The criterion can be thought of as a threshold for the expression of left-handedness. This could vary with cultural pressures against sinistrality. It is not possible to assess, on present evidence, whether the low incidences of sinistrality in some cultures are due to social pressures which move the threshold in a sinistral direction or whether they are due to high incidences of RS+ in some races. The series of dysphasics analysed below all derive from Western Europe and North America.

If individuals with right cerebral speech are RS− (except in cases of early left hemisphere pathology, excluded as far as possible from the data considered) the theory suggests that they would represent one half of the RS− group and be distributed over the whole range of relative manual skill. When the criterion is as drawn in Fig. 5, cutting off about 10 per cent of the total population, the percentage of left-handedness among those who are RS− is expected to be about 35 per cent. In the children of two left-handed parents I found 40 per cent sinistral writers. Thus among dysphasics with right unilateral lesions, it is expected that the *majority will be right-handed*. This prediction looks paradoxical since it is usual to emphasize the sinistral tendencies of right lesion dysphasics and their relatives. An excess of left handed relatives is predicted, of course, if RS− runs in families, but the suggestion that among the patients themselves, most would be right-handed prompted a re-examination of the main dysphasia series.

Test of a Theory

Table 1 shows the five large consecutive series of patients with unilateral cerebral lesions reported for handedness, lesion side and presence of dysphasia, assembled from the literature by Zangwill (1967) but substituting the revised British war wound series of Newcombe and Ratcliff (1973) for the earlier figures of Russell and Espir (1961). Conrad (1949) described the German war wound series. Penfield and Roberts (1959) reported patients operated for the relief of focal epilepsy, the figures here excluding patients with evidence of cerebral damage before the age of two years. Bingley (1958) studied patients operated for temporal lobe tumours. Hécaen and Ajuriaguerra's (1964) patients were a consecutive neurological clinic series

TABLE 1.   Incidences of dysphasia in five series of unilateral lesions analysed for handedness and lesion laterality.

Handedness		Left		Right	
Lesion		Left	Right	Left	Right
Conrad, 1949	N	19	18	338	249
	Dysphasic	10 (52·6%)	7 (38·9%)	175 (51·8%)	11 (4·4%)
Newcombe and	N	30	33	338	216
Ratcliffe, 1973		11 (36·7%)	8 (24·8%)	218 (56·2%)	19 (6·0%)
Penfield and	N	18	15	157	196
Roberts, 1959		13 (72·2%)	1 (6·7%)	115 (73·2%)	1 (0·5%)
Bingley, 1958	N	4	10	101	99
		2 (50·0%)	3 (30·0%)	68 (67·3%)	1 (1·0%)
Hécaen and	N	37	22	163	130
Ajuriaguerra, 1964		22 (59·5%)	11 (50·0%)	81 (49·7%)	0 (0·0%)

with varied unilateral pathology. The figures directly relevant to the prediction of more right- than left-handers among right lesion dysphasics are the *numbers* of patients (not percentages) in the right lesion groups. The first two series, both of war wounds, agree in finding more right- than left-handers. The next two series have very few cases in the categories of interest. The last series goes against the hypothesis with 0 right- and 11 left-handers.

The percentages of patients in the various groups vary widely between series, ranging from 36–72 per cent in the first column, 6–50 per cent in the second column and so on. In an attempt to discover just where the series differ they were compared for several variables (Annett, 1975). Table 2 shows the data collapsed to give the percentages in each series of patients with dysphasia, with left-sided lesion, and with left-handedness. (Combinations of these variables were also examined but need not be considered here.) Chi square comparisons were made of the numbers of cases in each category over the several series and very *few* significant differences were found. The percentages of dysphasics are remarkably similar in all series. The proportion of left-sided lesions was slightly smaller in Penfield and Roberts' series (1959) but as these were operative lesions depending on the surgeon's selection of cases the smaller number of left lesions is not surprising. This series will not be considered further here. (It is compared with the other series more fully by Annett (1975).) The remaining differences concern incidences of left-handedness, a little smaller for Conrad (1949) and considerably larger for Hécaen and Ajuriaguerra (1964).

TABLE 2.   Comparison of five series.

Series	Total N	Dysphasia (per cent)	Left lesion (per cent)	Left handed (per cent)
Conrad, 1949	624	32·5	57·2	5·9*
Newcombe and Ratcliff, 1973	767	33·4	54·5	8·2
Penfield and Roberts, 1959	386	33·7	45·3*	8·5
Bingley, 1958	214	34·6	49·1	6·5
Hécaen and Ajuriaguerra, 1964	352	32·4	56·8	16·8*
Total	2343	33·2	53·6	8·8

The series of Conrad (1949), Newcombe and Ratcliff (1973) and Bingley (1958), which did not differ significantly among themselves, were combined to give a total of 1605 cases, of whom 533 were recorded as dysphasic, as shown in Table 3. Among the dysphasics there were 7·7 per cent left-handers, suggesting a criterion close to that drawn in Fig. 5, but a little further to the left. Among the

TABLE 3.   Combined data from three series: Conrad (1949), Newcombe and Ratcliff (1973) and Bingley (1958).

Handedness	Left		Right	
Lesion	Left	Right	Left	Right
Total 1605	53	61	827	664
Dysphasic 533 (33·2%)	23	18	461	31
Among dysphasics				
Left handed	41 (7·7%)			
Right hemisphere lesion	49 (9·2%)			
Right hemisphere lesion and left-handed	18 (36·7%)			

dysphasics there were 49 (9·2 per cent) with right-sided lesions. On the assumption that these are half the RS− subgroup in the population, the population incidence would be about 18·4 per cent (drawn in Fig. 5 as 20 per cent). Among the 49 right lesion dysphasics there were 18 (37 per cent) left-handers, almost exactly as predicted.

Hécaen and Ajuriaguerra's (1964) series is analysed in Table 4. Among the 114 dysphasics there were 11 (9·7 per cent) with right hemisphere lesions, a proportion very similar to that of the combined series. Hécaen and Ajuriaguerra's data differ from the rest only in the criterion of left-handedness. Among the dysphasics, 33 (29 per cent) were recorded as left-handed. The incidence for the whole series was 16·8 per cent (Table 2) and for non-dysphasics the incidence was 11 per cent. Thus, the criterion of sinistrality differed markedly between dysphasics and non-dysphasics. It was shown

TABLE 4.   Hécaen and Ajuriaguerra (1964) series.

Handedness	Left		Right	
Lesion	Left	Right	Left	Right
Total 352	37	22	163	130
Dysphasic 114 (32·4%)	22	11	81	0
Among dysphasics				
Left handed	33 (28·9%)			
Right hemisphere lesion	11 (9·7%)			
Right hemisphere lesion and left-handed	11 (100%)			

above that the criterion of left-handedness could easily rise to about 35 per cent in normal samples and it is possible that more careful enquiries were made about the laterality of the patient and his relatives for dysphasics than for non-dysphasics. When about 29 per cent of the population are regarded as left-handed, the threshold in Fig. 5 must be drawn to the right of the line representing zero difference between the hands in skill. On this criterion, the majority of RS− individuals would be ascertained as left-handed, not right-handed as for the 10 per cent criterion. Thus, it can be inferred that the absence of dysphasic right-handers with right-sided lesions in this series is largely a function of the classification of handedness.

Hécaen and Piercy (1956) showed that in patients drawn from the same source as those of Hécaen and Ajuriaguerra, the incidence of dysphasia was very high in left-handers whichever the lesion side. They inferred that speech might be bilaterally represented in left-handers and this has been widely assumed to be the case in subsequent literature. The analysis above suggests that the large proportion of left-handers having dysphasic symptoms could be due to the inflation of left-handers among dysphasics and not to the inflation of dysphasics among left-handers. The present model suggests that bilateral speech would not be typical of individuals who are left-handed or of those who are RS−. It could occur in some RS− individuals who happen to be evenly balanced for cerebrality in the same way that a few truly ambidextrous individuals are evenly balanced for writing skill. Bilateral speakers would, as a group, have an excess of left-handed relatives, as would right cerebral speakers. The familial left-handedness is not a cause of the atypical speech, but rather another symptom of the lack of the right shift factor.

As the implications of this theory are examined in relation to further data, it may well be necessary to introduce complications. One obvious candidate for this role is gender. There are sex differences in the distribution of asymmetry of manual skill, in the incidence of developmental language problems and in the association of handedness between relatives. The sexes are not distinguished in the dysphasia data described above, but should be separated in future series. For the present it is worth emphasizing that the model makes a minimum of assumptions. It requires only one genetic factor and all the rest depends on chance.

References

ANNETT, M. (1964) A model of the inheritance of handedness and cerebral dominance, *Nature*, **294**, 59–60.

ANNETT, M. (1967) The binomial distribution of right, mixed and left handedness. *Quarterly Journal of Experimental Psychology*, **19**, 327–333.

ANNETT, M. (1970) A classification of hand preference by association analysis. *British Journal of Psychology*, **61**, 303–321.

ANNETT, M. (1972) The distribution of manual asymmetry. *British Journal of Psychology*, **63**, 343–358.

ANNETT, M. (1974) Handedness in the children of two left-handed parents. *British Journal of Psychology*, **65**, 129–131.

ANNETT, M. (1975) Hand preference and the laterality of cerebral speech. *Cortex*, **11**, 305–328.

ANNETT, M. (1976) A coordination of hand preference and skill replicated. *British Journal of Psychology*, **67**, 587–592.

ANNETT, M., LEE, D. and OUNSTED, C. O. (1961) Intellectual disabilities in relation to lateralised features of the EEG. In *Hemiplegic Cerebral Palsy in Children and Adults*, Little Club Clinics in Developmental Medicine, No. 4. London: Spastics Society/Heinemann.

BINGLEY, T. (1958) Mental symptoms in temporal lobe epilepsy and temporal lobe gliomas. *Acta Psychiatrica et Neurologica Scandinavica*, **33**, suppl. 120.

COLLINS, R. L. (1968) On the inheritance of handedness I. *Journal of Heredity*, **59**, 9–12.

COLLINS, R. L. (1969) On the inheritance of handedness II. *Journal of Heredity*, **60**, 117–119.

CONRAD, K. (1949) Cited by Zangwill (1967).

HÉCAEN, H. and AJURIAGUERRA, J. (1964) *Left-handedness*. New York: Grune & Stratton.

HÉCAEN, H. and PIERCY, M. (1956) Paroxysmal dysphasia and the problem of cerebral dominance. *Journal of Neurology, Neurosurgery and Psychiatry*, **19**, 194–201.

HOADLEY, M. F. and PEARSON, K. (1929) Measurement of internal diameter of skull in relation to "pre-eminence" of left hemisphere. *Biometrika*, **21**, 94–123.

NEWCOMBE, F. and RATCLIFF, G. (1973) Handedness, speech lateralisation and ability. *Neuropsychologia*, **11**, 399–407.

PENFIELD, W. and ROBERTS, L. (1959) *Speech and Brain Mechanisms*. Princeton, New Jersey: Princeton University Press.

PETERSON, G. M. (1934) Mechanisms of handedness in the rat. *Comparative Psychology Monographs*, no. 46.

RIFE, D. C. (1950) An application of gene frequency analysis to the interpretation of data from twins. *Human Biology*, **22**, 136–145.

RUSSELL, W. R. and ESPIR, M. L. E. (1961) *Traumatic Aphasia*. London: Oxford University Press.

WOO, T. L. and PEARSON, K. (1927) Dextrality and sinistrality of hand and eye. *Biometrika*, **19**, 165–199.

ZANGWILL, O. L. (1967) Speech and the minor hemisphere. *Acta Neurologica et Psychiatrica Belgica*, **67**, 1013–1020.

ETHNIC STUDIES ON SWEAT GLAND COUNTS

AGATHA S. KNIP

Institute of Human Biology, State University at Utrecht, The Netherlands

BETWEEN 1960 and 1975 several reviews appeared on the sweat glands in man and other species (e.g. Weiner and Hellmann, 1960; Jenkinson, 1973; Montagna and Parakkal, 1974). As the sweat gland plays an important role in thermoregulatory processes in man, it is not surprising that the main aim of these reviews was directed to a discussion of the sweat gland in terms of a functioning entity. Although various studies on sweat gland counts have recently been published, determination of total maximum number and regional distribution of the eccrine sweat glands has received relatively little attention.

In this paper, I shall discuss this rather neglected aspect of the human eccrine sweat gland.

Methodology

One can say that the eccrine sweat glands crudely have been studied from two points of view: physiological ("active" number) and histological ("anatomical" number). In this survey as far as the first type of study is concerned the maximum number of active glands is always meant (Knip, 1975).

As early as 1961 Weiner recognized the need to examine a sufficiently large sample of the body surface area in order to give reliable estimates of total number and regional distribution of sweat glands. Different authors (see Kuno, 1956) sampled from as little as 1 to about 12 cm^2 body surface area. It was recommended that a minimum body surface area of about 90 cm^2 be sampled in order to obtain a representative estimate of the total number (Weiner and Lourie, 1969).

Since the nineteenth century various techniques have been used to stimulate and to count the glands. Maximal activation mostly is achieved by drug- or thermal (local thermal)-induced stimulation while as counting procedures the starch–iodine and plastic impression techniques are frequently used (Collins *et al.*, 1959; Weiner and Lourie, 1969; Harris *et al.*, 1972). Table 1 shows the number of

TABLE 1.   Number of functioning sweat glands per cm^2 (F.S.G./cm^2) on the forearm of Caucasian subjects.

Author	Stimulus	Technique	Sample	F.S.G./cm^2
MALES				
Collins *et al.* (1959)	Drug	P.I.	4	71–109
Collins *et al.* (1959)	Local thermal	P.I.	4	78–111
Knip (1969)	Local thermal	P.I.	9	85·6[*]
Ojikutu (1965)	Local thermal	P.I.	74	105·8
Huebner *et al.* (1966)	Drug	S.I.	12	207
Randall (1946)	Drug	S.I.	4	213[*]
Randall (1946)	Local thermal	S.I.	3	212–250[*]
Sato and Dobson (1970)	Exercise	S.I.	14	108
Silver *et al.* (1964)	Drug	S.I.	7	152
Kawahata (1950)[**]	None	Droplet	6	149
Spruit and Reyen (1972)	Drug	P.I.[***]	1	300–320
Willis *et al.* (1973)	None	P.I.[***]	5	118–237
Krause (1844)	—	Anatomical	—	158[**]
Oberste-Lehn (1964)	—	Anatomical	121	152[****]
Szabó (1962)	—	Anatomical	3–21	220
FEMALES				
Knip (1969)	Local thermal	P.I.	6	93·4[*]
Huebner *et al.* (1966)	Drug	S.I.	13	207
Sargent and Weinman (1966)	Drug	S.I.	9	120
Sargent and Weinman (1966)	Local thermal	S.I.	7	120–130
Sargent *et al.* (1965)	Exercise	S.I.	4	150
Silver *et al.* (1964)	Drug	S.I.	8	199
Bar-Or *et al.* (1968)	None	S.I.	10	100–120
Oberste-Lehn (1964)	—	Anatomical	102	161·9[****]

[*]   extensor site
[**]   value derived from Marples (1965)
[***]   Silicone impression
[****]   age group 0–19 yr. included

P.I.: plastic impression	(Thomson and Sutarman, 1953)
S.I.: starch iodine	(Randall, 1946)
Local thermal: hot water bath	(Weiner and Lourie, 1969)
Drug: mostly methacholine	(Collins *et al.*, 1959)

active sweat glands expressed per cm² on the forearm of Caucasian subjects obtained with different techniques by different workers. When allowance is made for possible differences in site and size this picture illuminates the following points:

(1) Variation in numbers is to a great extent due to differences in technique.

(2) Different techniques used by the same worker yield more uniform results.

(3) Drug-induced stimulation gives similar numbers as compared to thermal (local thermal)-induced stimulation.

Recently, Harris *et al.* (1972) in their evaluation of different counting techniques proved the plastic impression method (Thomson and Sutarman, 1953) to be more sensitive than the starch–iodine method (Randall, 1946) but the ultimate choice seems to be the silicone impression method.

As was shown in the studies of Morimoto *et al.* (1967) and Bar-Or *et al.* (1968) another source of variation can be caused by submaximal activation.

Comparison between anatomical and active counts: do "inactive" glands exist?

The earliest figures of anatomical counts are those obtained by Krause (1844). Since then a few anatomical studies covering part or total of the human skin surface have been published (Thomson, 1954; Kuno, 1956; Szabó, 1962, 1967; Oberste-Lehn, 1964). From their results (especially Szabó's) it appeared that anatomical counts are appreciably higher than physiological counts in most studies.

This puzzling discrepancy may be due to technical errors such as distortion and shrinkage during fixation (Szabó) or to failure of enumeration methods (Harris *et al.*, 1972; Willis *et al.*, 1973). Another possibility may be a real discordance between the "active" and "anatomical" numbers of sweat glands. In fact, Ogata (1935) demonstrated the presence in the skin of eccrine sweat glands which showed no response to any form of stimulation and which were morphologically indistinguishable from active glands (Kawahata, 1950); the so-called "inactive" glands. According to Kuno (1956) the ratio of active to inactive varies from site to site and from subject to subject.

Some investigators (Collins *et al.*, 1965) doubt the existence of these glands when adequate thermal stimulus is applied. From the behaviour of the sweat glands during acclimatization, other workers (Ojikutu, 1965; Knip, 1975) presented indirect evidence in favour of their existence (see under acclimatization).

Undoubtedly, more specific studies are needed to elucidate this rather controversial but important issue.

Age, sex, body size and physique

The theory advanced by Kuno (1956) that eccrine sweat glands are not formed "*de novo*" after birth, has been supported by the studies of Szabó (1967) and Huebner *et al.* (1966). The observation by Oberste-Lehn (1964) that additional glands probably are formed only in the age group 40–60 year is based on insufficient evidence.

This proposition of Kuno means, that although the total number of sweat glands remains relatively constant throughout life, the density per unit area progressively decreases with growth (Thomson, 1954; Bar-Or *et al.*, 1968). Thus. an important source of variability in active counts seems to be caused by variations in body size.

Besides, Bar-Or *et al.* (1968) in their study on lean and obese subjects demonstrated a possible influence of physique on the number of active glands per unit surface area.

As far as the "thermal" eccrine sweat glands are concerned, the process of ageing itself seems to have no effect on the density in adult skin (Hellon and Lind, 1956; Szabó, 1962; Silver *et al.*, 1964; Giacometti, 1964).

Maximum total number of sweat glands shows quite similar values between men and women. Reported higher average densities in females can mostly be explained by their smaller body surface area (Kawahata, 1960; Bar-Or *et al.*, 1968; Szabó, 1967; Knip, 1969, 1972, 1975).

State of acclimatization

Kawahata (1950) showed in sixteen Japanese subjects of various ages that the total number of active glands is decidedly smaller in children under $2\frac{1}{2}$ years, whereas in the subjects above this age the number was higher and not related to age. Another study demonstrated a striking increase in total count in fifteen Japanese born in the tropics compared with four who migrated to the tropics after childhood. Also, the number in the tropical-born was rather similar to that obtained in ten Philippino natives. Besides, the total number in the Ainus living in a cool climate is reported to be approximately half that of natives of the tropics. To explain these facts it was postulated firstly that activation can only take place during the first $2\frac{1}{2}$ years of life, and secondly that acclimatization does occur but is definitely restricted to this age period. In other words it seemed that only in this age period and dependent on the climatic circumstances a larger or smaller number of "inactive" glands can be put into action.

In contrast, however, Thomson (1954) could find no difference in total number of active glands in adult Europeans and West Africans studied in Nigeria, and Weiner (1964) found similar sweat gland densities in Europeans and Bantus studied in South Africa.

It is conceivable that the observed enhanced sweat output during acclimatization to heat can also be caused by an increase (recruitment) in the number of active glands, e.g. "inactive" glands can be made to respond during this process. Most reports on the behaviour of the active gland during acclimatization failed to observe an increase in the maximum number of active sweat glands (Collins et al., 1965; Collins and Weiner, 1965; Peter and Wyndham, 1966; Sargent et al., 1965), although for the greater part these results are obtained by means of artificial acclimatization procedures.

In this respect the study of Ojikutu (1965) is of particular relevance. He observed much higher total sweat gland counts in an adult male African sample of Nigerians studied in West Africa as compared to a male African sample of Ghanaians and Liberians studied in Europe, where they had resided from one to five years. It was inferred that the Africans living in West Germany have lost part of their presumed tropical acclimatization.

This aroused my interest in the possible influence of post-infancy acclimatization on the maximum number of active glands. Adult males and females of Dutch and Indian (Hindu) ancestry were studied in the Netherlands (temperate climate) and in Surinam (hot humid climate). In addition, in Surinam a small sample of male Bushnegroes was included. Unfortunately, it was not possible to design the study with the same subjects moving and residing between the two climates.

The Dutch in Surinam showed distinctly higher total numbers than the Dutch in the Netherlands, but the Hindu showed similar total numbers in both climates, although the value was somewhat higher in Surinam. In the Netherlands no difference in total number was observed between the two groups. In Surinam the Bushnegroes were intermediate between the other two groups.

Analysis of the results suggests that the maximum number of functioning sweat glands changes during natural acclimatization. It cannot be excluded that the relative magnitude of this change depends on the percentage of individuals with the "recruitment type" of sweat glands within each ethnic group. Also, this change was not apparent in all body areas studied (Knip, 1975).

Although these conclusions need support from carefully designed studies on different ethnic groups it seems that some of the earlier conflicting results can be explained. For instance, the results of Kawahata, often criticized for the small body surface area studied (Collins *et al.*, 1965), also can be explained from the point of view of ethnic differences in percentage recruitment adapters. This argument was suggested by Johnson *et al.* (1969) in order to explain the results of Sargent *et al.* (1965) on sweat gland numbers during acclimatization in four females. Our finding that the trunk sites appeared to be the most resistant to climatic changes may explain the results of Peter and Wyndham (1966) obtained on the back and chest of Africans. Finally, the constancy of the maximum number of active glands on the forearm in Indians during natural de- and artificial re-acclimatization (Collins and Weiner, 1965; Collins *et al.*, 1965) corresponds with our finding in Hindu studied in Surinam and in the Netherlands respectively.

In conclusion, the state of acclimatization cannot be ignored in any study on maximum number of active eccrine sweat glands.

Do racial differences in active sweat gland counts exist?

In the older literature it is often stated that tropical races have more sweat glands than Europeans. In most cases, however, the evidence is not convincing. Thus, Clark and Lhamon (1917) studied the emotionally reacting palmar and plantar sweat glands; Schiefferdecker (1917) based his conclusions on apocrine glands, and Glaser (1934) studied only one Negro and one European.

Anatomical studies on Australian natives and Europeans (Green, 1971) and on corpses of different races (Kuno, 1956) indicated great individual variations, but real racial differences were not observed.

When allowance is made for non-genetic components as discussed

TABLE 2. Total number and average density of functioning sweat glands in male subjects of various ethnic groups.

Author	Ethnic group	Environment	Total number (millions)	Average density (F.S.G./cm²)	Body surface area (m²)
Austin and Ghesquiere (1973)	Bantu	Zaire, Africa (hot and humid)	1·38	(83·0)*	(1·66)
	Pygmoid		1·30	(88·9)	(1·46)
Ojikutu (1965)	Nigerian	Nigeria, Africa (hot and humid)	—	176·2	—
Thomson (1954)	Nigerian	Nigeria, Africa (hot and humid)	(2·20)	130·0	1·66
	European (imm.?)**		(2·20)	127·9	1·75
Weiner (1964)	Bantu	South Africa (subtropical)	<2	—	—
	European (imm.?)		<2	—	—
Ojikutu (1965)	European	Germany, Europe (temperate)	—	96·4	—
	African (imm.)		—	97·8	—
	Syr.-Iranian (imm.)		—	96·5	—
	Am. Negro (imm.)		—	83·2	—
Collins and Weiner (1965)	European	England, Europe (temperate)	<2	—	—
	Indian (imm.)		<2	—	—
Knip (1969, 1972)	European	Netherlands, Europe (temperate)	1·47	76·4	1·93
	Indian (imm.)		1·51	89·1	1·65
Knip (1974, 1975)	European	Surinam, S. America (hot and humid)	1·90	100·6	1·89
	Indian		1·56	86·8	1·81
	Bushnegro		1·69	95·0	1·79
Roberts et al. (1970)	Caingang Indian	Brazil, S. America (hot and humid)	(1·43)	(91·7)	1·56
Kawahata (1960)	Amer. White	North America	2·47	131	1·88
	Amer. Negro		2·18	117	1·86
Kawahata (1950)	Japanese	Japan	2·21	—	—
Kawahata (1950)	Japanese (imm.)	Davao, Philipines (tropical)	2·17	—	—
	Japanese		2·96	—	—
	Philipino		2·80	—	—
Kawahata and Sakamoto (1952)	Ainu	Hokkaido, Japan (cold)	1·44	—	—

* Figures in brackets are estimates   ** means immigrant (imm.)

TABLE 3.   Regional distribution of functioning sweat glands per cm² in various male ethnic groups.*

	Ojikutu (1965)			Roberts et al. (1970)	Knip (1972)			Ogata (1935)	Kawahata (1950)	Szabó (1967)
	African	Amer. Negro	Syrian/ Iranian	Caingang Indians	Dutch	Hindu	Bushnegro	Korean	Philipino	White (anatomical)
Forehead	352·7	158·9	137·7	—	113·4	120·4	198·3	290	420	360
Cheek	—	—	—	94	57·1	78·1	95·5	—	206	320
Chest	151·5	80·5	88·2	84	71·6	84·1	72·1	115	198	175
Abdomen	149·7	71·2	85·2	70	91·6	91·8	105·20	125	198	190
Scapula	174·7	73·0	94·3	61	65·5	86·9	78·8	130	211	—
Lumbar region	173·7	67·3	79·8	125	72·8	86·8	82·2	175	—	160
Arm	114·8	73·3	112·8	116	79·5	93·1	75·5	115	203	150
Forearm	201·4	79·5	100·3	—	85·6	121·3	113·0	140	231	225
Hand (dorsum)	—	—	—	—	145·1	170·4	191·0	215	284	—
Buttock	98·5	61·6	72·7	—	58·9	59·3	73·4	—	169	—
Thigh (medial)	—	—	—	—	37·4	49·4	52·4	—	—	—
Thigh (anterior)	—	—	—	—	56·9	63·8	64·9	—	172**	120**
Thigh (lateral)	—	—	—	—	62·3	66·0	75·3	—	—	—
Leg (medial)	—	—	—	—	64·2	78·3	87·9	145	166**	150**
Leg (lateral)	—	—	—	—	67·6	84·1	97·4	110	—	—
Foot (dorsum)	—	—	—	—	118·6	119·3	120·0	175	183	250

* Areas are not always directly comparable.
** Exact area unknown.

before, the existence of racial differences in physiological counts is also questionable. In Table 2 the total numbers of active glands and/or the average densities as reported in various adult male ethnic groups studied are tabulated. In general, different ethnic groups studied by the same author with the same method in the same environment yield rather similar results (Thomson, 1954; Ojikutu, 1965; Knip, 1969, 1972, 1974; Austin and Ghesquiere, 1973). However the possibility of adapting to changes in climate by a change in maximum number of active sweat glands may have a genetic basis. Our finding in Hindu and Dutch respectively may be a case in point.

The absolute figures vary to a great extent, as is illustrated by the values in the various African groups studied in Africa. Here, variations in degree of acclimatization may be partly responsible for the variation.

Table 3 presents data on the regional distribution in various ethnic groups. Nearly all series where sufficient sites have been examined show the typical pattern of distribution first described by Weiner (1964): a gradient on the limbs with the distal parts having more glands than the proximal. However, the "gradient" on the trunk with the lower part showing more glands than the upper part is not so clear. Analysis of variance indicated a tendency for sweat glands to have a different regional distribution in Europeans and Africans (Thomson, 1954; Ojikutu, 1965) but in view of the high inter-individual variability this observation needs confirmation.

Conclusion

It seems clear from this survey that non-genetic factors are more important than genetic factors as underlying causes for the observed variability in human eccrine sweat gland counts. However, methodology and state of acclimatization need more attention in future studies, and information on cold-adapted populations is lacking.

References

AUSTIN, D. M. and GHESQUIERE, J. (1973) Heat stress and heat tolerance in two African populations. Paper presented at symposium: "Biocultural adaption to environmental stress"; 72nd Annual Meeting Amer. Anthrop. Assoc. (unpublished).

BAR-OR, O., LUNDEGREN, H. M., MAGNUSSON, L. I. and BUSKIRK, E. R. (1968) Distribution of heat-activated sweat glands in obese and lean men and women. *Hum. Biol.*, **40**, 235–248.

CLARK, E. and LHAMON, R. H. (1917) Observations on the sweat glands of tropical and northern races. *The Anatomical Record*, **12**, 139–147.

122 PHYSIOLOGICAL VARIATION AND ITS GENETIC BASIS

COLLINS, K. J., CROCKFORD, G. W. and WEINER, J. S. (1965) Sweat-gland training by drugs and thermal stress. *Arch. environ. Health*, **11**, 407–422.

COLLINS, K. J., SARGENT, F. and WEINER, J. S. (1959) Excitation and depression of eccrine sweat glands by acetylcholine, acetyl-β-methylcholine and adrenaline. *J. Physiol.*, **148**, 592–614.

COLLINS, K. J. and WEINER, J. S. (1965) The effect of heat acclimatization on the activity and numbers of sweat glands: a study on Indians and Europeans. *J. Physiol.*, **177**, 16–17 P.

GIACOMETTI, L. (1964) The anatomy of the human scalp In *Advances in Biology of Skin*, vol. VI, edited by W. Montagna, pp. 97–120. Pergamon Press, Oxford.

GLASER, S. (1934) Sweat glands in the Negro and the European. *Am. J. phys. Anthrop.*, **18**, 371–376.

GREEN, L. M. A. (1971) The distribution of eccrine sweat glands of Australian Aborigines. *Aust. J. Derm.*, **12**, 143–147.

HARRIS, D. R., POLK, B. F. and WILLIS, I. (1972) Evaluating sweat gland activity with imprint techniques. *J. invest. Derm.*, **58**, 78–84.

HELLON, R. F. and LIND, A. R. (1956) Observations on the activity of sweat glands with special reference to the influence of ageing. *J. Physiol.*, **133**, 132–144.

HUEBNER, D. E., LOBECK, C. C. and MCSHERRY, N. R. (1966) Density and secretory activity of eccrine sweat glands in patients with cystic fibrosis and in healthy controls. *Pediatrics*, **38**, 613–618.

JENKINSON, D. McEwan (1973) Comparative physiology of sweating. *Br. J. Derm.*, **88**, 397–406.

JOHNSON, B. B., JOHNSON, R. E. and SARGENT, F. II. (1969) Sodium and chloride in eccrine sweat of men and women during training with acetyl-β-methylcholine. *J. invest. Derm.*, **53**, 116–121.

KAWAHATA, A. (1950) Studies on the function of human sweat organs. *J. Mie med. Coll.*, **1**, 25–41.

KAWAHATA, A. (1960) Sex differences in sweating. In *Essential Problems in Climatic Physiology*, edited by H. Yoshimura, pp. 169–184. Nankodo Publ. Co., Kyoto.

KAWAHATA, A. and SAKAMOTO, H. (1952) Some observations on sweating of the Aino. *Jap. J. Physiol.*, **2**, 166–169.

KNIP, A. S. (1969) Measurement and regional distribution of functioning eccrine sweat glands in male and female Caucasians. *Hum. Biol.*, **41**, 380–387.

KNIP, A. S. (1972) Quantitative considerations on functioning eccrine sweat glands in male and female migrant Hindus from Surinam. *Proc. Kon. Ned. Akad. Wetensch.* **C75**, 1, 44–54.

KNIP, A. S. (1974) Quantitative considerations on functioning eccrine sweat glands in male Bushnegroes from Surinam. *Proc. Kon. Ned. Akad. Wetensch.* **C77**, 1, 29–38.

KNIP, A. S. (1975) Acclimatization and maximum number of functioning sweat glands: a study on Hindu and Dutch females and males. *Ann. Hum. Biol.*, **2**, 261–277.

KRAUSE, K. F. T. (1844) Die Haut. In *Wagners Handwörterbuch der Physiologie*, vol. 2, pp. 126–131. F. Vieweg und Sohn, Brunswick.

KUNO, Y. (1956) *Human Perspiration*, pp. 1–416. C. C. Thomas, Springfield, Illinois.

MARPLES, M. J. (1965) Cutaneous appendages: the sweat gland. In *The Ecology of the Human Skin*, edited by M. J. Marples, pp. 22–32, C. C. Thomas, Springfield, Illinois.

MONTAGNA, W. and PARAKKAL, P. F. (1974) Eccrine sweat glands. In *The Structure and Function of Skin*, edited by W. Montagna and P. F. Parakkal, pp. 366–411, 433. 3rd ed. Academic Press, New York and London.

MORIMOTO, T., SLABOCHOVA, Z., NAMAN, R. K. and SARGENT, F. II. (1967) Sex differences in physiological reactions to thermal stress. *J. appl. Physiol.*, **22**, 526–532.

OBERSTE-LEHN, H. (1964) Effects of aging on the papillary body of the hair follicles and on the eccrine sweat glands. In *Advances in Biology of Skin*, vol. II, edited by W. Montagna, pp. 17–34. Pergamon Press, Oxford.

OGATA, K. (1935) Functional variations in the human sweat glands, with remarks upon the regional difference of the amount of sweat. *J. Orient. Med.*, **23**, 98–101.

OJIKUTU, R. O. (1965) Die Rolle von Hautpigment und Schweissdrüsen in der Klima-anpassung des Menschen. *Homo*, **16**, 77–95.

PETER, J. and WYNDHAM, C. H. (1966) Activity of the human eccrine sweat gland during exercise in a hot humid environment before and after acclimatization. *J. Physiol.*, **187**, 583–594.

RANDALL, W. C. (1964) Quantitation and regional distribution of sweat glands in man. *J. clin. Invest.*, **25**, 761–767.

ROBERTS, D. F., SALZANO, F. M. and WILLSON, J. O. C. (1970) Active sweat gland distribution in Caingang Indians. *Am. J. phys. Anthrop.*, **32**, 395–400.

SARGENT, F. II., SMITH, C. R. and BATTERTON, D. L. (1965) Eccrine sweat gland activity in heat acclimation. *Int. J. Biometeor.*, **9**, 229–231.

SARGENT, F. II. and WEINMAN, K. P. (1966) Eccrine sweat gland activity during the menstrual cycle. *J. appl. Physiol.*, **21**, 1685–1687.

SATO, K. and DOBSON, R. L. (1970) Regional and individual variations in the function of the human eccrine sweat gland. *J. invest. Derm.*, **51**, 443–449.

SCHIEFFERDECKER, P. (1917) Die Hautdrüsen der Menschen und der Säugetiere, ihre biologische und rassenanatomische Bedeutung, sowie die Muscularis sexualis. *Biol. Zentralbl.*, **37**, 534–562.

SILVER, A., MONTAGNA, W. and KARACAN, I. (1964) Age and sex differences in spontaneous adrenergic and cholinergic human sweating. *J. invest. Derm.*, **43**, 255–265.

SPRUIT, D. and REYNEN, A. Th.A. (1972) Pattern of sweat gland activity on the forearm after pharmacologic stimulation. *Acta Derm. (Stockholm)*, **52**, 129–135.

SZABÓ, G. (1962) The number of active sweat glands in human skin. In *Advances in Biology of Skin*, vol. III, edited by W. Montagna, R. A. Ellis and A. F. Silver, pp. 1–5. Pergamon Press, Oxford.

SZABÓ, G. (1967) The regional anatomy of the human integument with special reference to the distribution of hair follicles, sweat glands and melanocytes. *Phil. Trans. R. Soc.* Ser. B., **252**, 447–485.

THOMSON, M. L. (1954) A comparison between the number and distribution of functioning eccrine sweat glands in Europeans and Africans. *J. Physiol.*, **123**, 225–233.

THOMSON, M. L. and SUTARMAN (1953) The identification and enumeration of active sweat glands in man from plastic impression of the skin. *Trans. R. Soc. Trop. Med. Hyg.*, **47**, 412–417.

WEINER, J. S. and HELLMANN, K. (1960) The sweat gland. *Biol. Reviews*, **35**, 141–186.

WEINER, J. S. (1961) The counting of human sweat glands. *J. Anat.*, **95**, 451–452.

WEINER, J. S. (1964) In *Human Biology, an Introduction to Human Evolution, Variation and Growth*, edited by G. A. Harrison, J. S. Weiner, J. M. Tanner and N. A. Barnicot, pp. 401–506. Clarendon Press, Oxford.

WEINER, J. S. and LOURIE, J. A. (1969) *Human Biology, a Guide to Field Methods*, I.B.P. Handbook, vol. 9, pp. 427–441. Blackwell, Oxford.

WILLIS, J., HARRIS, D. R. and MORETZ, W. (1973) Normal and abnormal variations in eccrine sweat gland distribution. *J. invest. Derm.*, **60**, 98–103.

VARIATION IN SWEATING

J. S. WEINER

MRC Environmental Physiology Unit,
London School of Hygiene and Tropical Medicine

THIS paper is concerned with a number of factors which determine the extent of the inter- and intra-population variability in the sweat output under conditions of heat load. Of the population variability observed, it will be shown that a portion is a direct consequence of the test procedure itself, that some of the variability reflects differences in body size of the subjects and some is attributable to the subjects' differing history of heat exposure. Finally to be considered is the extent to which genetic differences may also enter into the observed variability.

Standardized Heat Load Tests

If a group of individuals, even of the same sex and ethnic group and of restricted age range, is exposed to a standardized heat stress the resultant sweat output shows a wide range of variation. Table 1

TABLE 1. Coefficients of variation in sweat output (Young European males "unacclimatized"; first test)

Standardized test	n	mean: ml/min	Coefficient of variation %	Reference
"Naval" Series N(O) climatic chamber	32	7·08	15·7	Hellon et al. 1956, p. 563
"Army" Series A climatic chamber	17	8·50	13·4	Fox et al. 1962, Fig. 1 and p. 101, Table 31, UTA; and 1969, Fig. 1
"Mining" Series M climatic chamber	17	8·38	20·4	Wyndham et al. 1965, Fig. 1
"Bed test" Series B	24	7·3	12·3	Fox et al. 1969, p. 451

gives illustrative data obtained from four different standardized tests widely used and referred to repeatedly in this paper. The mean sweat rates are fairly comparable but there is a wide range in the coefficient of variability.

Table 2 shows the conditions of testing in the three climatic chamber series, N, A and M listed in Table 1.

TABLE 2.   Climatic chamber conditions (Four hours exposure)

	D.B. °C	W.B. °C	A.M. cm/s	(A.M.) (ft/min)	Work	Water drunk
Series N(O)	38	29	50	(100)	1 ft 12 × min	fixed
Series A	40	32	25	(50)	1 ft 12 × min	fixed
Series M	34	32	25–40	(50–80)	height adjusted to give 1560 ft lb/min 12 × min	"ad libitum"

In these three climatic chamber series the subjects (wearing a minimum of clothing) were exposed to the thermal conditions for 4 hours. During this period they carried out step-climbing work at a rate of 12 steps per min, in 4 half-hourly bouts with intervening rest pauses. As Table 2 indicates, the three series differ in various ways. In none of the series, and more particularly in N and A, do the standardized conditions impose a uniform or equal heat load on the subjects. There are several reasons for this.

Firstly, there are demonstrable individual differences in heat gain at the skin surface by convection and radiation. This is because subjects vary in surface areas and in surface temperatures. Secondly (in series N and A), the work intensity and hence the metabolic heat gain from the work differs for the different subjects since no allowance was made for the variation in body weight in step-climbing. This was done in Series M. Thirdly, the bodily gain of heat (or "storage") varies between the subjects, partly because of variation in body mass and partly because of variation in mean temperature increase. It will be noted that body size variation enters into all three factors contributing to intra-group variability in heat load and hence to the variance in the resultant sweat output.

In the controlled hyperthermia test introduced by Fox, heat production by work is eliminated altogether. The subject is enclosed in a thin PVC vapour barrier (from within which sweat is collected by suction) and is wrapped within the air distribution layer, insulated by blankets, an aluminized sheet and a quilt. The head is enclosed in a heavily insulated hood. The core (ear) temperature is raised by introducing hot air until the target temperature of 38·0°C is reached. This level of hyperthermia is maintained for (usually) 1 hour. (For details see Fox in Weiner and Lourie, *Human Biology: a Guide to Field Methods*, 1969). The maximum sweat output occurs over the first 15 minutes ("initial" sweat) and this value is used in comparisons of groups in various stages of acclimatization. The initial sweat rate (Table 1—"Bed test" Series B) is independent of body weight or surface area.

In the chamber tests the four channels of heat flux, convection (*C*), radiation (*R*), metabolism (*M*) and storage (*S*), together account for the evaporative heat loss (*E*), and hence the sweat rate (less "drippage") in accordance with the steady-state heat balance equation

$$E = \pm C \pm R + M - S. \tag{1}$$

The variation in body size, that is, body weight and skin surface area, influences all four channels of heat flow.

Variance of Heat Gain at the Surface

Heat gain by convection (*C*) and by radiation (*R*) are given by

$$C = K_C A_C \sqrt{V}(T_A - T_S) \tag{2}$$

and

$$R = K_R A_R (T_R - T_S) \tag{3}$$

where

T_S is mean skin temperature for the exposure period
T_A is air temperature
T_R is mean radiant temperature
A_C and A_R are surface and radiant body surface areas
K_C and K_R are convective and radiative transfer coefficients respectively.

Bearing in mind that T_S at the start of heat exposure lies within the range 31°C to 34°C (Gagge, Herrington and Winslow, 1937; Fox *et al.*, 1969), it is clear that in series N and series A there will be a net flow of heat to the skin since T_A (and T_R) are above skin temperature. In the N series, temperature gradient will range from about 2°C to 4°C and in the A series from about 2°C to 6°C. There are two sources of variance. The variance in heat gain will be compounded by the variation in T_S and in A_C (and A_R), since surface areas range from about 1·6 to 2·1 m².

In the M series, where $T_A = 34$°C, convective and radiation avenues of heat gain are largely eliminated, and indeed there would be some loss of heat since during the heat exposure T_S would be of the order of 35°C to 36°C. Skin temperatures were not reported for this series.

Variance of Heat Gain from Work

In series N and A a second and major source of variance arises from the variation in body weight, which ranged from 55 to 85 kg. As all subjects were required to step-climb to the same height (1 ft) and at the same rate, the work performed is proportional to body weight (a man of 55 kg would do 1450 ft lb/min, one of 85 kg some 2246 ft lb/min). The coefficient of variation of body weight is of the order of 12 per cent and this is paralleled in the observed variation of energy expenditure for the combined step-climbing and resting routine, for which the coefficient of variation is 14 per cent (Fox, Jack, Kidd and Rosenbaum, 1962). Here again, series M contrasts with series N and A. Variation in work load was eliminated by adjusting the step-climbing height according to body weight so that all subjects worked at 1560 ft lb/min.

Heat storage

In the unacclimatized state there is a marked range in the level of core and skin temperature attained by the subjects in all the tests. This variation in temperature rise due to heat load is compounded by variation in body mass (and no doubt body water content) to produce variation in heat storage.

$$S = \Delta t_m \times W \times s \qquad (4)$$

where

Δt_m is increase in mean temperature of the body over the exposure period

W is body weight

s is specific heat (taken as 0·9).

The sources of variation in heat gain to the body in the chamber tests would, in the steady state, contribute to the variation in evaporative heat loss in accordance with the following linear function

$$\sigma_E^2 = \sigma_C^2 + \sigma_R^2 + \sigma_M^2 + \sigma_S^2 + \text{covariance.} \qquad (5)$$

For series N, A and M where $T_A = T_R$ and A_R can be taken as 0·8 A_C the expression becomes

$$\sigma_E^2 = \sigma_H^2 + \sigma_M^2 + \sigma_S^2 + 2r_{MH}\sigma_M\sigma_H - 2r_{MS}\sigma_M\sigma_S - 2r_{HS}\sigma_H\sigma_S. \qquad (6)$$

For the three series the prime data for resolving this function are not immediately available. Relevant data for a series of tests comparable to series N (O) but carried out on naturally acclimatized men in Singapore (series N VI, Macpherson, 1960, Ch. 5) can be used for illustrative purposes.

It can be shown for this series, in accordance with formula (6) that

(a) variance in heat production associated with step-climbing accounts for at least 70 per cent of the variance in total heat gain;

(b) variance in body weight accounts for about 80 per cent of the variance in heat production;

(c) body weight (through its correlation with surface area) accounts for some 10 per cent in variance of heat gain at the surface, and about 20 per cent of variance in storage.

Thus body weight variation contributes about 60 per cent of the variance in evaporative heat loss.

In this particular series the relation between sweat loss and body weight gives a correlation coefficient f of 0·76 (significant at the 0·1 per cent level), that is variation in body weight is associated with 58 per cent of the variance in sweat output. Since sweat output comprises both evaporated and non-evaporated sweat ("drippage"), this implies that·variation in the latter is also fairly deeply correlated with body size.

Two important sources of variation in sweat output (and in drippage), other than body size, arise from (i) intensity of habitual sweat gland usage or training, and (ii) level of water drinking during sweating.

Sweat Gland Usage

Post natal sweating

The investigations of Foster, Hey and Katz (1969) have established the existence not only of marked differences in sweat gland responsiveness soon after birth but also of large individual differences in the rate with which responsiveness to sudorific agents increases with age within 2 weeks post natally (Fig. 1).

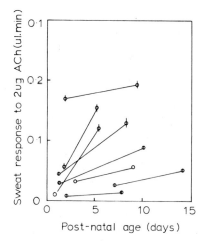

FIG. 1.   The effect of post-natal age on the maximum rate of sweat secretion after an injection of 2 μg ACh in 0·2 ml. saline. Results obtained in eight infants born within 2 weeks of term who were studied twice under comparable conditions in the first 2 weeks after birth. The vertical bars indicate the S.E. of the means (○) where $n \geqslant 4$. (From Foster, Hey and Katz, 1969.)

Foster *et al.* have also shown that in the adult living in temperate climates sweat gland activity, again tested by injection of mecholyl, has increased by threefold compared to the post natal value, as judged by sweat secretion per activated sweat gland or by duration of response. The actual amount of sweat secreted is nearly 7 times greater in the adult than in the newborn. There is about a 10 per cent coefficient of variation in responsiveness amongst the adults to this test. That an important component in bringing about variation in sweat gland activity is acquired through sweat gland use or training, and is not of genetic causation, is evidenced from the results of "acclimatization" procedures.

Acclimatization procedures: training of sweat glands

The fact that sweat glands can be made to respond more intensively with repeated stimuli is well brought out in experiments where mecholyl was injected daily in the same area (Fig. 2). Within the small sample of subjects there are quite marked individual differences in the rate of change in responsiveness.

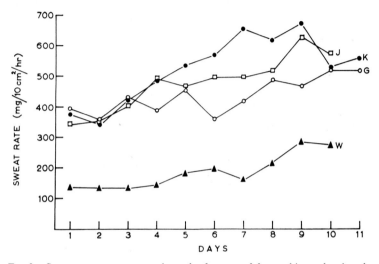

FIG 2. Sweat responses measured on the forearm of four subjects showing the improvement in sweating capacity induced by daily intradermal injections of methacholine.

The variation in sweat output observed in trials such as those of the A, M, N or Bed series (Table 1) can to some extent plausibly be attributed to the degree to which the glands have been used habitually. The way in which use and disuse of the glands can alter their responsiveness is well illustrated in a report by Collins and Weiner (1964). Eight Indian "Army" subjects were tested by methacholine injections on the first day of arrival in the UK in October. They were retested in November and December. As Fig. 3 shows, the sweat response on arrival in their naturally acclimatized state was consistently higher than subsequent responses when they were becoming unacclimatized as the colder weather set in. These subjects were then artificially acclimatized by ten days intensive exposure in the test room. This resulted in a markedly enhanced response to methacholine, above that of their "natural" level of acclimatization.

During these three phases, the eight subjects were also given repeated "uniformity" tests in the hot chamber, similar to the "A" series. The changes in sweat output paralleled those elicited by methacholine.

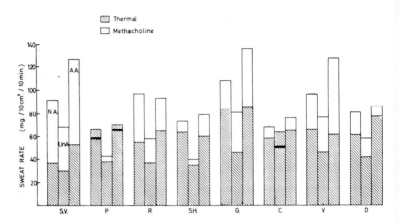

FIG. 3.   Histograms illustrating individual sweat responses of eight Indian subjects tested by heat and methacholine injections in the naturally acclimatized (NA), unacclimatized (UnA) and artificially acclimatized (AA) condition. The total body thermal sweat response is expressed in the same units as the forearm methacholine response. The total column represents the methacholine sweat rate and the hatched portion the thermal response. In only 3 of the 24 comparisons did the thermal exceed the methacholine sweat rate.

Variation and acclimatization

When subjects are given repeated daily exposures to a heat load as, for example, in the case of the M series on European and Bantu males by Wyndham *et al.* (1964), there is a noticeable reduction in the coefficient of variance as the sweat rate is increased by this artificial acclimatization procedure. The same phenomenon was found in the investigation of Hellon, Jones, Macpherson and Weiner (1956). Both the variance and coefficient of variance were reduced in the naturally acclimatized group in Singapore (series N) compared to the matched unacclimatized group in Oxford. Comparisons by the Fox test shows this same change quite consistently.

For example, in the study by Fox *et al.* (1969) the coefficient of variation was 58 per cent in a group before acclimatization and 34 per cent afterwards. An exception is reported in the paper by Fox *et al.* (1962) for subjects of series A also showed no reduction in variance with acclimatization. The reasons for this remain unclear.

If it is the case, as the majority of studies indicate, that a period of uniform treatment brings about a reduction in intergroup variance (as in series N), of about 25 per cent, this would be ascribable to a non-genetic factor independent of body weight. If this finding is applicable to the post natally acquired variability seen in unevenly and poorly acclimatized individuals, in the first tests by procedures such as M, N or A, then differences in body weight and previous experience would account for some 85 per cent of the variability. Thus some 15 per cent of the variability might be attributable to genetic factors (other than the genetic moiety associated with body weight differences). There is however another "non-genetic" factor which influences the level of sweat output, namely, water intake.

Water intake

In tests as long as 4 hours with high sweat output it is generally accepted that level of water intake can influence the sweat output. For example, in one study an intake of 550 ml/h produced a mean sweat output some 7 per cent lower than an intake of 850 ml/h (Macpherson, 1960, Table 101). For the series shown by Table 1, in series A and N fixed amounts of water were drunk after each water period whereas in series M "water was supplied *ad libitum*" and "subjects were encouraged to try to drink a sufficient volume of water to prevent weight loss". It is possible therefore that in series M sweat output may have been affected by variations in water drunk but the extent cannot be evaluated.

Ethnic comparisons

A number of studies have pointed to the finding in hot countries that in comparable tests of heat stress acclimatized Europeans generally show a greater sweat capacity than the technically indigenous groups. Evidence for the wide variety of racial groups who appear to differ from the European has been summarized by Fox *et al.* (1974):

"The sweat rates of the (local) Europeans on Karkar Island (New Guinea) were much higher than those of the New Guinea indigenes and closely comparable with the level for artificially heat-acclimatized Europeans. The male New Guineans had sweat rates similar to the unacclimatized Europeans, whereas the females in New Guinea were somewhat higher than their European counterparts. The low sweating capacity of the New Guinea people is in agreement with studies using the same function test on the indigenes of other hot countries such as Israel (Fox et al., 1973) and Nigeria (Ojikutu et al., 1972). It also agrees with the results of studies using the work-in-the-heat technique to compare Europeans and African people living in the same hot climate. These studies include the comparison by Robinson et al. (1941) of white and negro recruits to the U.S. Army by Baker (1958); a comparison of Europeans and Nigerians by Ladell (1950, 1957); several studies on the South African Bantu (Weiner, 1950; Wyndham, Bouwer, Devine and Patterson, 1952; Strydom and Wyndham, 1963), and a comparison of Chaamba Arabs with French servicemen living in the hot, dry climate of the Sahara desert (Wyndham, Metz and Munro, 1964). The relatively few studies on other races have generally given similar findings, thus the Australian Aborigine was found to sweat less than the white Australian by Wyndham and Munro (1964), although the converse has been reported by MacFarlane (1969).

Tests on indigenes in Singapore using the P4SR Index (McArdle et al., 1947) showed a lower sweat loss than that predicted for acclimatized Europeans (Macpherson, 1960). Naturally acclimatized Indians tested immediately after being flown from India have been found to have the same sweat rate as the unacclimatized European and when tested again after becoming de-acclimatized by living 3 months in England without exposure to heat, the Indians sweated significantly less than the Europeans (Edholm et al., 1965)."

Caution is called for in interpreting these results. Clearly in the ethnic comparison summarized by Fox it is not always certain to what degree the groups are strictly comparable in terms of body size, heat exposure experiences, and in water drinking. In some comparisons of pre- and post-acclimatization responses there has been a change in the number of subjects (e.g. Wyndham et al., 1964).

It is known that individuals can be acclimatized to a very high degree depending on the heat load and given sufficient exposure. Ojikutu *et al.* (1972) found a group of Nigerian workers in hot heavy industry who gave a mean "initial" sweat loss of 16 ml/min, which points to a high degree of "natural" or occupational acclimatization when compared to a value of 18·4 for a group of British males deliberately acclimatized.

The number of strictly controlled comparisons of different ethnic groups after equivalent periods of deliberate acclimatization is quite small. One cannot say more than that they confirm the impression that Europeans in the well acclimatized state tend to sweat more than similarly treated indigenes.

This does not necessarily indicate that a genetic difference is involved. The fact that groups as genetically diverse as Chinese, Malays, Australian Aborigines, Negroes and Indians may sweat more "economically" than Europeans seems as much an argument for some peculiar European genotype as for some environmental factor operating in the acquisition of heat tolerance by Europeans.

An explanation along these lines is afforded by the observations of Macpherson (1960, Chap. 10) on water drinking patterns. These observations have shown clearly that the increased sweating of the artificially acclimatized state is associated with a correspondingly high water intake. But after 4 or 5 months of heat exposure both water intake and sweat rate begin to fall, though they remain above the pre-acclimatization test level. If indigenes or those long habituated to living conditions in the heat practice an economy of water intake through restriction on their drinking habits, or perhaps because they experience some reduction in thirst sensitivity, this would be accompanied by a reduction in sweat wastage at high sweat losses. The order of difference in sweat output so far reported between heat acclimatized Europeans and other groups is not so large as not to be compatible with the effects of reduced water drinking.

The Heritability Component

On our present information, it would be premature to ascribe ethnic differences in sweat output to genetic differences. Even if intrapopulation sweat variation, as suggested above, is ascribable to a heritable factor to the extent of, say, 15 per cent it would still not follow that a similar difference between groups would also be demonstrable. (This point has been cogently argued in the case of inter- and intra-group comparisons of IQ scores by Owentin, Bodmer and others).

Nevertheless, it remains of clear interest to attempt to determine more precisely whether the within group genetic variability in sweating contributes perhaps no more than 15 per cent of the observed variability. Twin studies would help to provide this information.

Only one twin study appears to have been carried out so far (by P. Clark, R. H. Fox, W. E. Glover, A. J. Hackett, R. J. Walsh and P. Woodward). This is an investigation, not yet published, on an Australian sample predominantly in the age range 17 to 25, consisting of 17 female monozygotic and 16 dizygotic pairs, 11 male monozygotic and only 1 male dizygotic pairs. The variation in age and sex, and the unevenness of the twin categories makes interpretation difficult. The subjects all underwent the controlled hyperthermia test. A preliminary analysis of concordance (which Dr. R. H. Fox has kindly allowed me to quote) appears to show very little indication of a genetically ascribable difference in sweat output.

This study would seem to support the view that the genetic determination of sweat response variations is small as compared to the effects of such factors as sweat gland training and usage and water drinking.

References

Collins, K. J. and Weiner, J. S. (1964) The effect of heat acclimatization on the activity and numbers of sweat-glands: a study on Indians and Europeans. *J. Physiol.*, **177**, 16–17P.

Foster, K. G., Hey, E. N. and Katz, G. (1969) The response of the sweat glands of the new-born baby to thermal stimuli and to intradermal acetylcholine. *J. Physiol.*, **203**, 13–29.

Fox, R. H., Budd, G. M., Woodward, P. M., Hackett, A. J. and Hendrie, A. L. (1974) A study of temperature regulation in New Guinea people. *Phil. Trans. R. Soc. Lond. B*, **268**, 375–391.

Fox, R. H., Jack, J. W., Kidd, D. J. and Rosenbaum, S. (1962) *Acclimatization to Heat.* Army Personnel Research Committee, Medical Research Council, London, Report APRC 61/25, pp. 31–98.

Fox, R. H., Löfstedt, B. E., Woodward, P. M., Eriksson, E. and Werkstrom, B. (1969) Comparison of thermoregulatory function in men and women. *J. appl. Physiol.*, **26**, 444–453.

Gagge, A. P., Herrington, L. P. and Windslow, C-E. A. (1937) Thermal interchanges between the human body and its atmospheric environment. *Amer. J. Hyg.*, **26**, 84–102.

HELLON, R. F., JONES, R. M., MACPHERSON, R. K. and WEINER, J. S. (1956) Natural and artificial acclimatisation to heat. *J. Physiol.*, **132**, 559–576.

MACPHERSON, R. M. (1960) *Physiological Responses to Hot Environments.* Medical Research Council Special Report Series No. 298. London: H.M.S.O.

OJIKUTU, R. O., FOX, R. H., DAVIES, T. W. and DAVIES, C. T. M. (1972) Heat and exercise tolerance of rural and urban groups in Nigeria. *Proc. Conf. Human Biology of Environmental Change, Blantyre, Malawi* (ed. D. J. M. Vorster), pp. 132–144. England: Gresham Press.

WEINER, J. S. and LOURIE, J. A. (1969) *Human Biology: a Guide to Field Methods* (*IBP*). Oxford: Blackwell Scientific Publications.

WYNDHAM, C. H., MORRISON, J. F. and WILLIAMS, C. G. (1965) Heat reactions of male and female caucasians. *J. appl. Physiol.*, **20**, 357–364.

WYNDHAM, C. H., STRYDOM, N. B., MORRISON, J. F., WILLIAMS, C. G., BREDELL, G. A. G., VON RAHDEN, M. J. E., HOLDSWORTH, L. D., VAN GRAAN, C. H., VAN RENSBURG, A. J. and MUNRO, A. (1964) Heat reactions of Caucasians and Bantu in South Africa. *J. appl. Physiol.*, **19**, 598–606.

A MULTINATIONAL ANDEAN GENETIC AND HEALTH PROGRAMME: A STUDY OF ADAPTATION TO THE HYPOXIA OF ALTITUDE

WILLIAM J. SCHULL

Center for Demographic and Population Genetics
University of Texas Health Science Center
Houston, Texas

and FRANCISCO ROTHHAMMER

Department of Cellular Biology and Genetics
University of Chile
Santiago, Chile

TWELVE thousand years ago, possibly earlier, man appeared in the highlands of South América near Ayacucho (MacNeish, Berger, and Protsch, 1970). Assuming the prehistoric migrants to the Americas arrived over the Bering land bridge, the descendants of some had made their way surprisingly rapidly to South America and to one of the more inhospitable regions of the New World, an area which differs strikingly in oxygen tension, humidity and temperature from most of the land through which they must have passed. We know virtually nothing of the physical appearance of these individuals, and are not even certain whether they merely frequented the altiplano in search of game—only their tools have been found, which might suggest this to be so—or resided there permanently. We do know, however, that individuals accustomed to the levels of oxygen which obtain at or near sea level when transported to this altitude are apt to experience one or some combination of the following: shortness of breath, dizziness and marked palpitation

139

upon exertion, exhilaration, sleeplessness, headache, the irregularity of respiratory rhythm known as Cheyne–Stokes breathing, impairment of mentation, diminished night visual acuity, accelerated pulse as well as other less obvious signs (Buskirk, 1971). Ultimately, however, within days or weeks at most, an accommodation is made, and one's normal activity level can again be achieved without undue physical stress or apprehension. The process by which this occurs is termed acclimatization to distinguish it from the adjustment which proceeds over generations and is commonly known as adaptation. Conspicuous changes associated with acclimatization are an increase in haemoglobin and circulating red blood cells, and to a smaller degree behaviour modification, a tendency to hyperventilate and initially at least to husband energy.

In the twelve millenia which have intervened since his arrival, Andean man has obviously thrived; more than ten million individuals now live in the Andes at altitudes of 3000 m or more. Indeed, communities exist at elevations of 4500 m, and active unaided mining at almost 6000 m has been reported (Keys, 1936; Buskirk, 1971). This increase in numbers undoubtedly reflects an ability, at least in part, to influence environmental circumstances through the domestication of a wide variety of frost-resistant plants and the discovery of means to store surpluses in the form of herds of alpaca and llama, for example. But has man himself adapted? Has his genetic make-up been systematically altered as a consequence of the rigours of this environment?

Nineteenth century observers (e.g., Forbes, 1870) noted that the Aymara and Quechua indigenous to Bolivia, Chile, and Peru differed anthropometrically one from another but had in common tremendous chests which they correctly ascribed to the demands of the altitude. Forbes (1870; p. 208), for example, asserted that

> The great size of the body of the Aymara Indian, which compared with his other dimensions, cannot fail to attract immediate attention; and a closer examination at once shows that of this the major part is occupied by the region of the chest ...

These observers were also intrigued by the changes in bodily dimensions seen in lowland Aymara colonists; Forbes (1870; p. 220) stated,

> These Indians, besides being as a rule somewhat taller men, appear

to have lost very much of their massive build, and become more slender and flexible in their forms and movements . . .

Stewart (1973) apparently concludes from such evidence that man's adjustment to altitude, as seen in these indigenous peoples, is not transmitted, and is not, therefore, a true adaptation—a conclusion which seems to us premature. But what evidence does, indeed, exist?

Andean man differs from his lowland counterparts in a number of anatomical, biochemical, and physiological particulars (Hurtado, 1964, 1971; Monge and Monge, 1966; Frisancho, 1975). These include (1) a prominent chest with long sternum and large thoracic capacity (Forbes, 1870); (2) an increased frequency of ventilation (Hurtado, 1964); (3) polycythaemia, increased reticulocytes, and a generally increased haematopoiesis (Merino, 1949); (4) hyperplasia of the marrow, increased iron utilization and turnover (Reynafarje, Lorans and Valdivieso, 1959; Reynafarje, 1966); (5) high haematocrits; (6) elevated serum uric acid; (7) more extracellular fluid; (8) slow pulse; and (9) a low systemic blood pressure but often a mild degree of pulmonary hypertension associated with increased pulmonary vascular resistance ostensibly secondary to the structural characteristics of the pulmonary arterial bed (Arias–Stella and Saldana, 1963; Penaloza, Sime, Banchero, Gamboa, Cruz and Marticorena, 1963). There are, in addition, (10) cardiovascular and electrocardiographic changes, such as in ventricular depolarization which are alleged to be virtually unique (Penaloza, Gamboa, Marticorena, Echevarria, Dyer and Gutierrez, 1961). Changes have also been reported (11) in levels of 2,3-diphosphoglycerate encountered among the Quechua (Lenfant, Torrance, English, Finch, Reynafarje, Ramos and Faura, 1968) and in the Bohr effect (Morpugo, Battaglia, Bernini, Paolucci and Modiano, 1970), that is, the decrease in oxygen affinity of haemoglobin when the pH falls. Whether the two observations under (11) are independent, and indeed, whether both effects exist has occasioned some controversy (DeBruin, Janssen and Van Os, 1971). They should, however, both lead to more ready tissue release of oxygen at lower partial pressures (Chanutin and Curnish, 1967; Benesch and Benesch, 1967), an obviously useful adjustment. Finally, (12) Reynafarje (1966) has called attention to differences in fructose metabolism between high altitude natives and sea level residents, and tissue oxidative enzymes which suggest important changes in tissue utilization of oxygen.

Functional adaptation to high altitude hypoxia, as Frisancho (1975) has noted, could occur through modifications in (1) pulmonary ventilation, (2) lung volume and pulmonary diffusing capacity, (3) transportation of oxygen in the blood, (4) diffusion of oxygen from blood to tissues, (5) utilization of oxygen at the tissue level, or some combination thereof. All of these alternatives may be operative but differentially. It would not be unexpected that the genetic contributions to these various adaptive pathways might differ; environmental modulation of growth and development should loom large in some pathways and small in others.

While a compelling argument can be made that many, if not all, of the findings we have cited are adaptive, whether they are in fact so remains open to challenge. We know very little about the genetic contribution to any of the changes we have previously enumerated save, possibly 2,3-diphosphoglycerate. Variation in the latter may be multigenic (Brewer, 1974), and sexes and races are known to differ (Eaton and Brewer, 1968); as yet, little variation has been found among individuals in respect of the enzyme 2,3-diphosphoglycerate mutase, although at least one variant of the "normal" form has been described and deficiencies are known to occur (Chen, Anderson and Giblett, 1971; Schroter, 1965; Vives-Carons, Kahn, Hakim and Boivin, 1973). Proof of the adaptive value of the observed changes seems to entail not only the demonstration that these variables change systematically with changes in the selective force, here oxygen tension, but that, in fact, a significant component of interindividual variation at any altitude stems from genetic variability. Much of the genetic contribution to the adaptation of the contemporary Aymara and Quechua must have occurred generations, quite possibly millenia ago, and thus is inaccessible to study. However, at a number of locations in Bolivia, Chile and Peru, the recent past has seen a migration of these people to lower altitudes. This downward, frequently coastward, movement, affords an opportunity to study changes in variability which may be indicative of a loss in adaptation, a deadaptation or relaxation of selection as it were, which may be related to the genetic contribution to a past, largely historic process. Rigour can be ascribed to such arguments only if there exists a measurable, biologically significant genetic contribution to the observed variation among individuals. Indeed, it is also necessary to demonstrate that reproductive differences can be associated with phenotypic ones, that is, that there exist differences in Darwinian fitness. Ideally, the evidence for genetic modulation of the variables thought to have adaptive value would stem from populations indigenous to hypoxic areas, rather than be inferred from lowland cultures, but this may be an unrealistic stricture.

Much of the clinical and physiological evidence that relates to man's adaptation to the hypoxia of altitude may be misleading. Our data are too often based upon occupationally selected, physically fit, young-to-middle aged males. It is not clear how representative these individuals may be of the populations of which they are members. Is the Andean man, whose particulars we have previously sketched, really typical of all Aymara or Quechua males at altitude? How does Andean woman differ? Has this adaptation, if such it be, been without penalty? Are not the increased frequencies of congenital abnormalities of the heart, e.g., patent ductus arteriosus, and chronic diseases of the liver and gall bladder, evidence of the price paid for his and her adjustment to hypoxia (see Alzamora-Castro, Torra, Batillana, Abugatas, Rubio, Bournocle, Zapata, Santa Maria, Binder, Subiria, Pando, Paredes and Perez, 1952)? Systematic changes in temperature and humidity accompany the changes in oxygen tension associated with altitude. May not the effects ascribed to hypoxia confound those of these other physical parameters?

This presentation sets forth in detail a study, known as the Aymara Health Survey, (part of a broader Multinational Andean Genetic and Health Programme), which aims to measure changes in these people potentially attributable to changing partial oxygen pressures (and accompanying changes in humidity and temperature). It is not only a description of a process but of a place and its people. This investigation also provides, we believe, a broader model for collaborative studies which address not only important scientific issues but pressing public health concerns as well—in this particular instance cardiovascular and pulmonary diseases which exhibit a localized or systemic pathologic hypoxia.

The land and the people

Much of the area immediately south of the present border of

Chile and Peru, the Departamento de Arica of the Provincia de Tarapaca, is a desert without vegetation, and with little or no rain. The shore is a pink-brown cliff face rising in some places to a height of 1000 m, pierced here and there by usually modest rivers which arise in the cordilleras to the east. At the bottom of this cliff are sea-eroded terraces, on which are built the towns. Beyond the coastal escarpment at an elevation of 600 m or so, and somewhat to the south, are the Pampas de Chaco and Camarones, a series of old lake floors of varying widths, the northern limits of the Atacama Desert. Eastward the land rises rapidly into the Sierra de Huaylillas and the Cordillera Central to fall briefly to the altiplano and then rise again to the Cordillera Oriental which forms the boundary between Bolivia and Chile; the crest of these mountains has an average elevation of 5000 m with occasional peaks over 6000 m high. This latter barrier is broken at about 4000 m by the Chapiquina Pass, some 80 km from the sea.

Scattered across the altiplano and through the quebrada of the Sierra de Huaylillas and the Cordillera Central are villages of Aymara (see Fig. 1 and Table 1). The 5000 to 10 000 Aymara who live in the area (many more are in Bolivia) are descendants of a once formidable Andean culture (La Barre, 1948; Murra, 1968). Their villages and caserios range from tens of individuals to a few hundred. Until recently they have been primarily peasant pastoralists deriving their livelihoods from their herds of llama and alpaca, or subsistence agriculturalists, the former in the altiplano and the latter in the sierras. Archaeological evidence suggests that the highlands of Peru have been inhabited by man for 10 000 to 12 000 years; some 3000 to 6000 years ago human beings were eking out a precarious existence in the coastal valleys and 1000 to 1500 years ago individuals, presumably the antecedents of the present inhabitants of the Arican sierra, were decorating rocky faces and ledges with simple figures of themselves, their animals and other fauna of the region (see Niemeyer Fernandez, 1972).

Five to six hundred years ago much of the coastal area was conquered by the Inca who moved some of their subjugated people into this region (Garcilasso de la Vega, 1604; Rowe, 1945). The origins of these forced colonists are not clear, but it seems unlikely that they were drawn from the altiplano (see Monge and Monge, 1966). This use of coastal colonists seems, however, to predate the

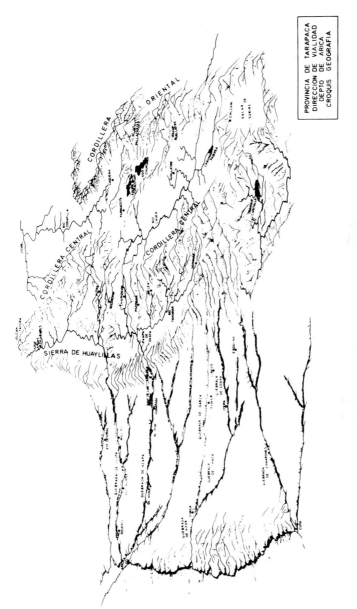

FIG. 1. A physiographic sketch of the Departamento de Arica with some of the village locations.

TABLE 1. The sizes of the various districts in which examinations occurred as recorded in the 1970 National Census, the *ad hoc* censuses initiated by this study, and the numbers of individuals examined

Distritos	1970 National Census	1973–1974 census	Persons examined	Participation rate (%)
Putre	408	419	381	90·9
Socoroma	160	159	88‡	55·3
Tignamar	288	249	135	54·2
Belen	325	233*	202	86·7
Lauca	186	129	89	69·0
Parinacota	278	357	245	68·6
General Lagos ⎫	538	466*	315	67·6
Cosapilla ⎭	295			
	833			
Azapa	3110	204† (San Miguel)	286 (29)	—
Poconchile ⎫	739	131† (Poconchile)	279 (66)	—
Molinos ⎭				

* Our censuses did not always embrace a full distrito. In Belen they include only the poblacion Belen, the caserio Chapiquina and the Campamento Central Chapiquina. In the distrito Cosapilla here combined with General Lagos, only the caserios Copatanga, Cosapilla, and Humaquilca along with the Lugar Agricola Guacolla were enumerated.
† No effort was made to enumerate the entire population of the Azapa, Poconchile, and Molinos distritos. These areas have been and are growing rapidly. To estimate their size we censused one caserio in the Azapa Valley, San Miguel, and one aldea, Poconchile, in the Lluta. Their present populations are about 360 and 110% of the size given in the 1970 census. If these values are representative of the communities in the valleys generally, then the Azapa Valley must presently have about 12 000 inhabitants and the Lluta Valley about 800. We will have seen, therefore, approximately 2·5% of the residents in the one valley and 25% of those in the other.
‡ Examinations in Socoroma had to be terminated earlier than expected because of the delay in the initiation of the survey in 1973 due to the political unrest in Chile in the autumn of that year. This accounts in part for the lower participation rate.

Incan conquest, and may have originated with the Aymara themselves (Murra, 1968). Within a hundred years of the Incan conquest, the Spanish arrived, and subdued the Inca and Aymara. An effort was made to concentrate the nomadic shepherds around Catholic churches in the altiplano and sierra, but with only limited success and by 1776 the Spanish had withdrawn from the higher altitudes, largely as a result of the exhaustion of the silver and gold mines. The only city of the sierras in Arica which dates from this general period is Belen; all of the other communities are almost exclusively Aymara and predate the Spanish conquest (Wormald, 1969). In the last several decades more of the Aymara have been moving from the altiplano to the sierran valleys, and even to the neighbourhood of Arica. Economic intercourse with the coastal Indians, is, however, much older; Aymara artifacts have been found in the gravesites in the Azapa (Safford, 1917; Bird, 1943), one of several major coastal lowland valleys. These valleys were inhabited before the Spanish occupation of Chile (see Bird, 1963).

Broadly speaking, the Departamento de Arica is divisible into three sharply differentiable but adjacent ecological zones—the coastal zone, the sierra and the altiplano, which differ strikingly in temperature, rainfall and oxygen pressure; associated with changes in these are biotic changes which limit the types of economies which can be practised, and thus life styles. We shall now briefly describe these different environments (see also Keller, 1946; Cruz-Coke, Cristoffanini, Aspillaga and Biancani, 1966; Santolaya, Donoso, Apud and Sanudo, 1973).

The coastal zone embraces four valleys, the Lluta, the Azapa, the Codpa and the Camarones, and the terrace on which the present city of Arica sits. While portions of these valleys are lush, as a result of irrigation and the rivers which arise in the mountains, rainfall is sparse, and this appears to have been true throughout their human occupation (Bird, 1943). Many coastal areas of the Departamento habitually see less than one millimetre of rain per year. Temperatures are moderate. The annual average maximum in Arica is 21·8°C and the annual average minimum is only 14·9°C; the diurnal cycle is small, less than 10°C. Seasonal differences are limited to a five-month cloudy period (June–October) when each morning the sky is possessed by low clouds which dissipate, during the day and most afternoons are sunny. It is a salubrious climate, moderated by the Pacific high pressure centre to the west, and swept by the southeast trade winds.

These are conditions conducive to crop diversity; grains, olives, grapes, melons, sugar cane, vegetables and a variety of citrus fruits are or have been grown. Even cotton was raised commercially at one time (Squier, 1878, p. 219). Agricultural enterprises, which can be initiated at any time of the year, now depend either upon labour imported from other areas in Chile or from the west of Bolivia.

Malaria was either already prevalent in the area, or became so shortly after the Spanish arrived, and persisted until 1945 (Wormald, 1968).

In the sierra, the terrain is rugged and highly irregular. Villages either squat on tablelands or commingle with highland valleys; most are at elevations of 3000 to 3500 m and are generally situated near springs or mountain streams. At these heights, partial oxygen pressure is approximately 100 mm of mercury (as contrasted with 149 mm at sea level). Rain, though markedly seasonal, occurs with increasing intensity from 2500 m upwards. Temperatures, particularly the daily and annual minima, are substantially lower than those of the coast, and the diurnal temperature cycle often exceeds 30°C. The soil is generally poor, and crops are less diverse; the most important one is alfalfa grown as fodder for the cattle and sheep. Oregano and some vegetables are cultivated for home use. Llama and alpaca, contrary to common thought, are not particularly numerous, and some wild guanaco persist. Trees are few, and then mostly imported conifers or eucalyptus. Irrigation is primitive but effective. Equatorially (north) facing slopes are more productive; presumably the difference of a few degrees in soil temperature is significant. Cosmic and ultraviolet activity are much more intense. In the dry season the humidity rarely reaches 20 per cent.

Over 40 per cent of the Departamento de Arica lies above 4000 m; much of this is a rolling high plain contained on the east and west by mountains which occasionally exceed 6000 m—the so-called altiplano. It is an austere land, alternately wet (December–March) and dry (April–November), punctuated in the north by salt flats and intimidated in the south by an active volcano, Guallatiri. Temperatures commonly oscillate as much as 30°C in the course of 24 hours; changes of as much as 50°C are not infrequent (Keller, 1946). Snowfall is slight, except in the high mountains, and is confined almost exclusively to the rainy season. Little shelter exists from the chilling, parching winds, creators of "dust devils" and rachitic plants, which blow through much of the year; there are virtually no trees, save the tortuous quenua (*Polylepis incana*), and the coarse grass and scrubby growth rarely attain a metre's height. Here and there are marshy oases of green, the bofedales, which support herds of alpaca and llama. Vicuña, the graceful kin of the alpaca and llama; viscacha, a shy rock-loving rabbit-like rodent; nandu, the South

American rhea; condors and perdiz, a partridge of a kind, are commonplace, and the lakes, Chungara and Cotacotani, are graced with pink flamingos, the parina, which rarely stray (Hernandez, 1970). Potatoes are staples; other crops are few. Corn will not thrive, and quinua, which does well in Bolivia, does poorly here. A large cauliflower-like woody plant, called *llareta*, provides one of the few combustibles with which to cook or relieve the cold. Skin weathered to an elephantine abrasiveness, cheeks often unnaturally rosy, and eyes etched with an actinic patchwork of capillaries characterize the human inhabitants of this area. This is a remarkable place and a remarkable people.

Objectives

Broadly stated, the objective of this study is to appraise the genetic contribution to man's adaptation to hypoxia. We seek, first, to evaluate the effects of environmental differences, particularly oxygen tension, upon pulmonary function, cardiopulmonary relationships and degenerative cardiac disease, such as arteriosclerosis, in the Aymara, and to assess the genetic contribution to their anatomical, biochemical and physiological responses.

Second, we aim to evaluate the burden of disease and disability prevailing in populations which reside at these altitudes and under these forms of agriculture and animal husbandry.

Third, we propose to measure in these people some of the normal constituents of blood (cholesterol, glucose and serum uric acid) which elsewhere are associated with an increased risk of coronary artery disease. The former two are prominent risk factors, and the third somewhat less so.

Fourth, through the assessment of the frequencies of a number of biochemical and immunological genetic markers, we expect to identify (*a*) the extent, if any, of non-indigenous contribution to the present gene pool, and (*b*) those simply inherited biochemical variants which appear to be responding directly, in frequency, to changes in oxygen tension. We expect the latter to be associated with red cell glycolysis, oxygen transport, or the regulation of the latter, and to be recognizable because their frequencies will correlate with changing oxygen pressure. The villages, are generally small, and thus stochastic effects might be erroneously identified as systematic ones. To avoid this, if possible, we measure a second array of genetic markers, not identified with glycolysis nor known to be involved with oxygen transport. This array is presumably not under systematic, selective pressure, but is responsive to the stochastic elements (drift and founder effects), which contribute to change in small populations, and to migration. Since members of more than one village have been measured in each of the three broad ecological niches under scrutiny, the "within village–within niche" mean square will serve as an estimate of the "error variance" against which to judge the significance and importance of apparently systematic changes in specific enzyme systems.

Finally, our fifth objective is to measure as many of the glycolytic intermediates involved in the regulation of oxygen transport as practicable. Our interest in enzymes, the glycolytic path and their products, rests on the supposition that the most likely avenue of genetic adaptation to hypoxia will be through those enzymatic pathways associated with oxygen transport or the tissue utilization of oxygen. These studies will embrace the kinetic characteristics of the enzymes as well as their structural forms as revealed by electrophoretic techniques, for it is conceivable that evolutionarily important changes in the kinetics could occur without structural changes recognizable by electrophoresis.

The examination

The examinations or observations include: (1) complete medical history and general physical examination, (2) oral and dental examinations, (3) anthropometric measurements, (4) pulmonary function tests, (5) resting electrocardiogram, (6) ophthalmoscopic examination including visual acuity, colour vision tests, and tensiometry, (7) detailed appraisal of cardiovascular status, (8) simple performance tests such as tapping, (9) nutritional, reproductive and residential histories, and finally (10) either an ACD-preserved specimen of venous blood, or a perchloric or trichloracetic acid precipitated specimen, or both.

To achieve as much standardization of the various procedures as practicable, manuals in English and Spanish were prepared which described in detail how each measurement was to be made, observations obtained and recorded, and the intent of each question on the various questionnaires. Training sessions and periodic discussions and evaluations of performance occurred in the field.

Enlistment of participants

Important objectives, in addition to the scientific goals, are to provide medical and dental care to the individual villager, to seek an evaluation of health problems in the villages of the interior, and, if possible, to make recommendations to guide the efforts of the Junta de Adelanto de Arica and the Servicio Nacional de Salud in improving the availability of health care in an area sparsely settled and difficult of access. Clearly, the success of this study hinged entirely upon our capacity to involve the individual villager and to satisfy his or her expectations wherever feasible.

The Aymara have often been characterized as dull, stolid and unimaginative (see, for example, Tschopik, 1963); these generalizations are inaccurate and unfair. They are perhaps an unsophisticated folk, and in this area of Chile (and western Bolivia) chary of outsiders, as experience has rightly taught them to be. Rapport is not easily established, but can be won if (*a*) they are candidly apprised of what their participation in a study entails, and (*b*) they clearly perceive some benefit to themselves. Aside from a few policlinics, manned by practicantes (carabineros trained to function as paramedical personnel), no medical attention is routinely available to them. Obviously they are concerned about their health and that of their children. They could, therefore, see benefit to the health examinations which comprise an integral part of our study. Most, however, are unfamiliar with such devices as a lung function analyser, an electrocardiograph, or an audiometer. To allay their concerns and to apprise them fully of what the examinations entailed the following routinely occurred:

(1) Before our appearance in a particular village, representatives of the Junta de Adelanto de Arica, familiar to the villagers, called upon the village authorities to explain our study and its objectives.

(2) Upon our arrival, but before the initiation of actual examinations, the villagers were invited to a public meeting at which the programme was carefully explained, and a film illustrating each of the successive steps in the investigation was shown. Questions prompted by either the film or the explanation of the programme were welcomed.

(3) When an individual presented himself or herself to be examined, the steps in the examination were again explained, and the

individual was assured that he or she was free to refuse any part of the examination process without prejudice to participation in the remainder, or to treatment or referral for treatment of their health problems where such treatment could be extended.

(4) Finally, upon completion of the examination, a physician explained to the participant any significant findings on his or her health, responded to any queries, dispensed any medication deemed necessary, and presented the individual with a written summary of the findings, and certain normative values, e.g., blood pressure, height, weight, etc. Where medical follow-up was indicated, or more extensive treatment required, arrangements were made with the Servicio Nacional de Salud hospital to have the person accepted for the treatment in Arica.

The examinees: Their number and locations

As originally envisaged, approximately equal numbers of individuals (and families) were to be examined in the coastal valley communities (0–300 m), the sierran villages (2500–3500 m), and the altiplano hamlets (4000 m and above). Most of the villages are accessible only by four-wheel drive vehicles. All of the examinations occurred either in October or November of 1973 or 1974. At this time of the year it is dry, roads are passable, and the temperatures more tolerable. Examinations were made either in schools or a social hall, or a village meeting place. Twelve different examination sites were used, one each in the Azapa and Lluta valleys, six in the sierras (Belen, Chapiquina, Murmuntani, Putre, Socoroma and Tignamar), and four in the altiplano (Caquena, Guallatiri, Parinacota and Visviri). The numbers reflect the population densities and the ease or difficulty of travel. In the coastal valleys, public transport was available, but in addition, we had vehicles to transport examinees from outlying villages to the examination locations. Transport was also provided in the sierra and altiplano, where walking is the only normal means of transport. Table 2 sets out the village of residence and the village of examination for the 2096 individuals who were examined; their ages ranged from 15 days to over 90 years (Table 3).

Censuses were conducted in the villages of the coast, sierra and altiplano before the initiation of examinations by employees of the Junta de Adelanto de Arica or village school teachers. Comparisons

TABLE 2.   The distribution of examinees by place of residence and place of
examination

Place of residence*	Region of examination			
	Coast	Sierra	Altiplano	Total
Coast				
Arica	33	30	3	66
Valle Azapa	286	0	0	286
Valle Lluta	279	0	0	279
Other	1	0	0	1
Sierra				
Belen	0	88	0	88
Chapiquina	0	113	1	114
Murmuntani	0	30	0	30
Putre	0	381	0	381
Socoroma	0	58	0	58
Tignamar	0	135	0	135
Altiplano				
Caquena	0	3	146	149
Charana	0	0	3	3
Guallatiri	0	0	89	89
Parinacota	0	0	96	96
Visviri	0	0	315	315
Unknown	4	1	1	6
Total	603	839	654	2096

* Residence here implies presently living in the stated village or its immediate environs;
boarding students are assigned to the village in which their school is and not to the
village of their parents if the two are different.

of the census findings with the rosters of the individuals actually
examined reveals our study population to represent 70 to 93 per cent
of any given village in the sierra and altiplano (see Table 1 for an
analysis of participation based upon distritos rather than villages);
no effort was made to study exhaustively the coastal communities.
Our resources were inadequate for the latter, and moreover most
of the inhabitants of these communities are not Aymara. Most of
the individuals who were not examined were elsewhere with their
flocks and herds, or lived too far from the village of census to make
an appearance at the examination site practicable. We are not
aware of any systematic bias that this may have introduced, nor is
such a bias suggested by the sex and age characteristics of the
"not-examined" group as indicated in the censused areas.

Some notion of the extensiveness of the family data which is

TABLE 3.   Distribution of the ages of the examinees by sex and region of examination

Age (yrs)	Village of examination							
	Coastal		Sierran		Altiplano		Total	
	Male	Female	Male	Female	Male	Female	Male	Female
<1	12	13	14	15	22	21	48	49
1–4	29	27	17	39	33	32	79	98
5–9	38	46	79	70	53	53	170	169
10–14	54	59	81	59	58	64	193	182
15–19	30	16*	40	18*	19	24	89	58
20–29	32	46	37	44	31	35	100	125
30–39	38	41	42	28	36	28	116	97
40–49	35	25	40	41	33	32	108	98
50–59	12	12	35	34	26	19	73	65
60–69	14	14	33	30	15	7	62	51
≥70	5	5	20	23	6	7	31	35
Total	299	304	438	401	332	322	1069	1027

* Note the deficiencies in these classes. We presume this reflects the commonly encountered reluctance of young women of these ages to disrobe for the physical examination and anthropometric measurements.

available can be gleaned from Table 4 where we distribute nuclear families as a function of the number of parents and offspring examined.

Ancestry and residence

Two matters of moment, basic to the study, are the algorithms by which an individual is designated, first, to be Aymara, and second, a coastal, sierran, or altiplano resident. Aymara family names are recognizably different from Spanish ones, and thus presumably individuals with such names are readily classifiable (for a listing of Aymara names with their frequencies see Table 5). Most Aymara in the Departamento de Arica are Catholics, at least nominally, and substantial numbers have taken or more frequently been given Spanish surnames at baptism, and such persons pose classificatory problems if the surname is the sole criterion of Aymara origin. Wormald (1969; p. 46) states that the census of 1871 reveals 39 to 57 per cent of the Aymara inhabitants of the precordilleran villages of Belen, Chapiquina, Putre and Socoroma had Spanish surnames, whereas 25 to 56 per cent of the Aymara inhabitants of

TABLE 4(a). The distribution of "nuclear" families examined as a function of the number of parents and offspring examined

Number of siblings examined	Number of parents examined			
	0	1	2	Total
1	833	142	40	1015
2	87	51	30	168
3	27	28	44	99
4	11	19	25	55
5	2	10	15	27
6	0	4	4	8
7	0	2	1	3
8	0	1	1	2
Total	960	257	160	1377

TABLE 4(b). The distribution of individuals examined based upon the number of their first degree (parents, offspring, siblings) relatives who were also examined

Number of relatives examined	0	1	2	3	4	5	6	7	8	9	10
Frequency	472	345	284	272	288	201	126	62	27	10	1

the altiplano villages of Caquena, Guallatiri and Parinacota were so named. The origins of some of these names are known, and frequently their advent in a village can be dated (Urzua, 1969).

As more roads have penetrated the sierra and altiplano, and as the economic attractions of the city of Arica have grown, largely in the last three decades, the population of the interior has become more mobile. Many individuals have moved permanently to the city; others visit the coast for days, weeks, or even years but consider their residence to be in the interior. No fully satisfactory method for the assignment of such individuals to residential groups exists. If one insists that an individual who is to be designated a resident of the altiplano must have been born at that altitude, and never lived at any other altitude for even a day, too few individuals will qualify. Similarly, if the strictures on residence are too loose, real differences provoked by a lifetime at different oxygen pressures may be obscured. We have selected a middle road.

Residence is here defined on the basis of (a) place of birth, and (b) proportion of life lived at a given altitude. Thus, an individual born in an altiplano village who has spent at least 90 per cent of his

TABLE 5. Enumeration and distribution of those family names encountered ten or more times in at least one of the three ecological niches of interest. Aymara and Quechua names are in italic.

	Altiplano	Sierra	Coast
Aguirre	—	8	18
Alanoca	2	15	2
Alave	93	21	—
Alberto*	51	1	—
Alcon	22	—	3
Alvarez	32	29	20
Ancare	—	—	10
Ancase	—	12	2
Anqueltoma	19	—	—
Apata	12	20	1
Apaz	32	50	1
Apaza	15	1	15
Ape	—	11	3
Arellano	—	1	91
Aros	—	10	1
Ayca	3	—	33
Baltazar*	25	7	1
Beltran	—	1	11
Benitez	—	14	3
Berna	3	13	—
Blanco*	61	76	33
Blas*	101	—	1
Branez	3	10	—
Bravo	1	1	18
Caballero*	14	23	2
Caceres*	2	38	28
Calizalla	18	11	4
Calle	52	94	40
Canque	1	20	3
Cariz	15	1	—
Carmona	—	—	12
Carrasco	8	27	19
Carvajal	—	5	11
Casas	14	4	—
Castillo	3	11	20
Castro	1	29	14
Chambe	26	39	5
Choque	56	141	54
Choquechambe	2	16	5
Chura	79	2	4
Churata	15	17	—
Colque	30	52	6
Conajagua	—	—	11
Conde	—	12	2

TABLE 5. (continued)

	Altiplano	Sierra	Coast
Condori	53	41	29
Contreras	—	18	20
Cortes	1	6	31
Crispin	—	14	—
Cruz*	45	3	19
Cutipa	—	22	2
Delgado	—	24	3
Diaz	1	16	12
Escobar	—	1	11
Espinoza	1	11	1
Fernandez	6	41	32
Flores*	144	135	48
Fuentes	—	—	10
Gallardo	—	1	10
Garcia	6	9	15
Garnica	11	9	1
Gomez	14	49	21
Gonzalez	1	17	21
Gregorio	—	—	16
Guajardo	2	—	10
Gutierrez	22	53	32
Henriquez	—	10	1
Herrera	19	4	57
Huanca	56	86	47
Huarache	1	40	66
Huaylla	71	2	—
Humire	22	32	30
Imana	14	—	—
Inquiltupa	70	2	4
Jaramillo	—	10	—
Jimenez	38	13	7
Jiron	—	10	9
Juarez	—	—	10
Lanchipa	5	—	12
Lara*	—	27	1
Larva*	—	16	1
Lazaro	6	55	—
Limari*	5	22	1
Llusco	21	2	—
Loayza	—	15	1

TABLE 5. (continued)

	Altiplano	Sierra	Coast
Lopez	17	6	23
Lovera	—	6	21
Loza*	1	33	—
Luque	4	29	2
Machaca	3	13	1
Maita	65	4	—
Mamani	173	261	215
Manlla	—	10	1
Manzano*	27	—	3
Marca	16	29	8
Mayorga	—	1	10
Medina*	19	47	11
Menacho	—	2	10
Mita	21	1	11
Mollo	26	50	21
Montes	10	—	—
Morales	28	28	7
Munoz	—	11	6
Nina	11	2	—
Nunez	3	4	17
Ocana	—	17	9
Olivares	—	—	10
Onofre	14	—	—
Orozco	4	15	2
Pacaje	55	30	3
Pacci	15	16	—
Paco	54	7	2
Pare	18	1	3
Paredes	4	10	—
Perez	8	30	42
Poma	87	29	14
Quilca	20	3	—
Quinones	—	14	1
Quispe	36	56	20
Ramirez	6	4	16
Ramos	1	30	19
Rivera	—	4	11
Robledo	—	—	19
Rodriguez	6	6	10
Rojas	4	6	17
Romero	11	6	1
Roque	26	—	—
Sanchez	31	30	3

TABLE 5. (continued)

	Altiplano	Sierra	Coast
Santos	2	10	1
Silvestre*	24	—	—
Solis	—	6	15
Tancara	39	15	5
Tapia	44	32	23
Tarque	9	16	13
Terraza	—	5	29
Ticlle	—	16	1
Ticona	26	10	2
Tito	36	5	3
Tolla	—	14	2
Trigo	—	—	11
Tupa	1	—	17
Valenzuela	—	1	14
Varas*	15	—	1
Vargas	7	3	24
Vasquez	2	82	32
Vega	—	4	10
Veliz	1	19	10
Ventura	—	13	—
Vicente	—	17	1
Vilca	18	43	25
Villalobos*	73	3	8
Villanueva	22	—	10
Villazon	5	11	—
Viza	12	14	—
Yampara	—	2	33
Yucra	78	48	23
Zanga	14	6	—
Zarate	—	—	14
Zarzuri	57	14	2
Zegarra	1	26	9
Zubieta	3	7	10
Sum	2759	2952	2015
Total names encountered†	2952	3527	2888
Total different names	204	355	488

* These names, though ostensibly Spanish, most commonly refer to individuals of Aymara ancestry when the name occurs in the sierra or the altiplano. Flores, for example, is an often encountered translation into Spanish of the Aymara name Pancara (or Tancara).
† Each individual examined contributes four and possibly six names to this enumeration, namely, a paternal patrinym and matrinym, a maternal patrinym and matrinym, and if married, the paternal and maternal patrinyms of the spouse.

or her subsequent life in an altiplano village will be considered an altiplano resident and assigned to his current village of residence. If, however, he or she has resided mostly at some lower altitude, assignment will be to his or her current village in that altitude. This procedure should be a conservative one in the sense that those effects, genetic or otherwise, secondary to one altitude will be assigned to a lower altitude and thus obtund the "true" differences which should prevail if no residential ambiguities existed. The effects of this definition are illustrated in Table 6 where we give the numbers of individuals born in the coastal, sierran and altiplano communities who assert that they have lived either their entire lives

TABLE 6.   Residential permanence as a function of residential niche

Residential niche	Percentage of life lived in stated niche							
	100	>90	>80	>70	>60	>50	<50	Total
Coast	391	5	15	19	22	34	145	631
Sierra	410	53	59	27	38	32	193	812
Altiplano	483	41	26	16	9	16	62	653
Total	1284	99	100	62	69	82	400	2096

at their present elevation (possibly in a different village, however), or for other amounts of time. Our study, though principally directed towards adaptation, can also provide insight into the process of acclimatization. In addition to the 2096 individuals under discussion we have also examined 429 persons living in the communities of Toledo and Turco in the Bolivian Department of Oruro. These individuals afford a contrast for the Bolivian Aymara migrants encountered in the coastal valleys of Arica, and a basis of comparison for the Chilean Aymara whose altiplano environment is more hostile and less productive than the areas immediately to the east of the cordillera where more rain falls.

Acquisition and management of data

Studies of this size and complexity require substantial preliminary preparation, not the least of which concerns the management of the data as they are collected. The traditional field notebook hardly suffices as a means of recording observations. More than 300 measurements (exclusive of laboratory determinations) were made on or questions asked of each participant, the average day saw 35 or more individuals examined, and never less than 12 examiners or interviewers were involved in the collection of the observations. To achieve the implied rate of data acquisition and the requisite uniformity among examiners and interviewers all questionnaires or observational records as far as possible had to be self-coding, and to manage conveniently the array of measurement or observational domains the records associated with each were colour-coded. All forms were bilingual, Spanish and English. Most Chilean Aymara understand and can speak Spanish; for those who did not, one of the examiners was also fluent in Aymara, Quechua and Uru-Chipaya, collectively the major languages of the altiplano of Bolivia, Chile and Peru. All forms were completed in duplicate.

Once the records were machine-retrievable, a "logical" verification of each occurred, that is, all coded values were examined to see whether they lay within the alternatives specified in the code itself, singly and as a logical group. Many questions, for example, were contingent upon the answer to a prior question; the logic of such concatenations was verified. All quantitative observations were screened singly and pairwise. In the first instance, means, variances, maxima and minima were determined to ascertain whether these accorded with experience or reason. In the second instance, all bivariate arrays were plotted by machine to search for "outliers" or discrepant observations. This search occurred within age class, altitude group, or for a specific measurement when multiple observations of a particular kind were made, e.g., as in the case of lung function analysis. All seemingly atypical measurements were re-examined (with the aid of the original form), and ultimately either the discrepancy was reconciled or the measurement was assigned to an "unknown" category.

Potentially of great importance were those data relevant to gene segregation, for a surprisingly small number of family studies have been done on the attributes associated with adjustments to altitude. Isolation of this subset of observations was not, however, straightforward. Many of these people are illiterate; moreover, there is no convention for the written forms of their surnames, if the latter are Aymara. Each interviewer could, therefore, transliterate a particular name differently. Families frequently did not appear at the examination location as a group, and even had this been so, it was likely that they would have been interviewed by different persons. Accordingly, it was necessary once the data were machine-retrievable to standardize names, that is, to verify that a given name was always transliterated in the same manner. This does not, of course, imply that our transliteration was phonetically the best, but merely that we were consistent. Consistency was necessary, for we proposed to link (and have linked) family units on the basis of their names (as well as subsidiary evidence). Linking of families through names is unquestionably more accurate in countries which follow the alleged Spanish tradition, that is, where an individual bears two surnames, a patrinym and a matrinym.

Finally, all atypical biochemical genotypes have been, or will be contrasted with as many "standards" as available or practicable. All have been, of course, examined in the context of the family in which they occur.

Data analysis

Extensive anthropological, biochemical, clinical, demographic, physiological and sociological observations are available. The analyses which could be envisaged border on the infinite. We have assigned priorities based partly on our perceptions of the timeliness of a particular analysis, and partly on the data retrieval and analytic difficulties involved. We propose first to undertake the analysis of the (1) growth and development, (2) blood pressure observations, (3) pulmonary function measurements, (4) physical examinations, (5) audiometric and ophthalmic data, and finally (6) the psychometric observations. This will leave numerous areas of analysis uninitiated; we propose, therefore, to turn next to analyses of (*a*) the genetic distance between villages and ecologic niches; this will include description of the frequencies of the genes associated with the various genetic markers both in the glycolytic path (or the pentose shunt) and those outside this metabolic pathway; (*b*) the estimation of genetic admixture which may exist; (*c*) the family data to ascertain the genetic contribution to variation within lineages in respect of the qualitative and quantitative adjustments to various levels of hypoxia; (*d*) the correlation structure which exists at the various altitudes in relation to growth and development, and pulmonary function; indeed, we are interested in the correlative behaviour between all pairs of measurement domains under scrutiny now; (e) the results of the assays of glycolytic intermediates, both insofar as our data may contribute to understanding of the genetic contribution of these metabolites, and to the role of the latter in the red cell's response to hypoxia. Other areas worthy of study include (f) the demographic and fertility data, and the relationship of variation in these to possible differences in gene frequency; and (g) patterns of migration, as revealed by the residential histories, and (h) the clinical findings.

Each of the areas enumerated (a)–(h) are substantial undertakings, but important since little information of this kind exists on populations of this nature. We have not attempted here to set out specific hypotheses or issues, for many seem self-evident. Thus, for example, thrombosis seemingly is a rare phenomenon among individuals who live at high altitudes, but whether this reflects some enzymatic adaptation to hypoxia or is artifactual in the sense that other causes of death intervene before the age at which thrombosis becomes common, is uncertain. Caen, Ergueta, Michel, Daufresne, Poupart and Dhuime (1971) have advanced some observations which suggest the former. They find that the effect of adenosine diphosphate (ADP) on platelet aggregation is less among Aymara and Quechua than among Europeans. Iatridis, Iatridis, Markidon and Regatz (1975) have recently shown that 2,3-DPG is a physiological inhibitor of platelet aggregation; this observation may account for Caen et al.'s finding since 2,3-DPG is elevated among Indians indigenous to high altitudes (Lenfant et al., 1968). We shall have data on ADP, ATP and DPG, which could be informative.

Similarly, it has been reported (see Frisancho, 1975) that mean blood pressure, systolic and diastolic, does not increase with age among high altitude Indians. The evidence for this is, however, somewhat tenuous for the sample sizes on which this observation rests were small and relatively few individuals of middle and later years were studied. Our observations literally extend over 90 years as previously indicated, and sample numbers are substantial from ages 6 to 60.

The family data, in both the nuclear and extended senses of family, provide a means to assess environmental and genetic contributions to variables such as blood pressure, cholesterol and growth variables in a variety of ways. Thus, we can calculate dyadic correlations between various classes of relatives, e.g., parent–offspring, sib–sib, or employ the family set strategy which has been advocated when one is interested in statements about environmental components of variation as little confounded with genetic variation as possible (Schull, Harburg, Erfurt and Rice, 1970; Chakraborty, Schull, Harburg and Schork, 1977).

The demographic observations, and in particular the data on reproductive performances can be used to quantitate reports of fertility differences at altitude (Donayer, 1966; McClung, 1969) and to relate performance to "biochemical types". For example, is an individual's prospects of survival during gestation and the first days or weeks of life dependent upon his or her mother's level of 2,3-DPG? Alternatively put, is fetal and infantile mortality at altitude related to the mother's capacity to unload oxygen at the tissue level; differences in genetic fitness are ultimately differences in reproductivity, either as a result of greater survivorship, greater fecundity, or both. Our data can also serve to support earlier observations that menarche is delayed at altitude (Donayer, 1966), and to inquire whether menopause is also altered.

Some preliminary findings

Three new, or at least different insights into the biology of the Aymara which have emerged thus far will be cited to illustrate our expectations of this study. These specific examples have been selected for the diversity they represent; we shall not enlarge here upon their interpretation. First, we note that significant differences in the frequency of systemic hypertension (that is, a systolic pressure of 160 mm Hg or greater, a diastolic pressure of 95 mm or greater, or both) exist between niches and ethnic groupings, as well as within an ethnic grouping in different niches. Hypertension is more common over all altitudes among adult non-Aymara (33/220), than Mestizos (28/405), and more common in the latter than among the Aymara (17/396). The frequency of hypertension increases with increasing altitude among adult non-Aymara, decreases among the Aymara, but remains more or less constant among the Mestizos.

Second, the routine use of mydriatics (1 per cent mydracil) instilled in a standardized manner to enhance visualization of the posterior chamber, fundus and retina revealed 42 individuals among 673 whose irides failed to dilate satisfactorily within 25 to 30 minutes. These individuals were dissimilarly frequent among the Aymara, Mestizo and non-Aymara groups which make up the sample of ophthalmic examinees. The phenomenon was most common among the Aymara (8·8 per cent), less common among the Mestizos (5·8 per cent), and the least common among the non-Aymara (2·7 per cent). Distribution of these individuals within the three groups suggests the trait to be inherited. Other pharmacogenetic traits are, of course, known in man, such as isoniazid inactivation and an inherited atropine esterase which degrades atropine rapidly occurs in rabbits and alters their response to mydriatics.

Finally, some 65 individuals among the approximately 1700 who have been studied electrophoretically for 15 enzyme systems have isozymic variants which appear new. The enzymes involved are glucose-6-phosphate dehydrogenase (G6PD), 6-phosphogluconate dehydrogenase (6PGD), phosphohexose isomerase (PHI), lactic dehydrogenase (LDH), and adenylate kinase (AK). All of these can be related directly or indirectly to glycolysis. Thus, glucose-6-phosphate dehydrogenase controls the flow of glucose into the hexose monophosphate shunt and away, as it were, from the early steps in glycolysis; under certain conditions, indeed, those associated with altitude adaptation (for example, elevated levels of 2,3-diphosphoglycerate), the flow of metabolites through this aerobic pathway may be controlled by 6-phosphogluconate dehydrogenase rather than G6PD. Phosphohexose isomerase appears to exert its influence on glycolysis through its metabolic interaction with hexokinase, the latter is the controlling enzyme for entry of glucose into glycolysis in normal erythrocytes. The action of lactic dehydrogenase is more indirect; the serum lactate : pyruvate ratio which is influenced by the action of LDH in turn influences the NAD : NADH ratio and can thereby affect the activity of glyceraldehyde-3-phosphate dehydrogenase by limiting the available NAD. Finally, adenylate kinase functions in the production of adenosine diphosphate (ADP) from adenosine monophosphate (AMP) generated through the purine salvage pathway. ADP is then further phosphorylated to adenosine triphosphate through the action of glycolysis. We do not presently know whether these enzymic variants are beneficial, prejudicial, or have no impact on the well-being of their possessors in an hypoxic environment. In view of the roles of these specific enzymes in red cell metabolism it would seem unlikely, however, that their alteration would not have some impact for better or worse.

Acknowledgments

The study we have described is both multidisciplinary and multi-national; we are privileged, therefore, to write on behalf of our colleagues in Bolivia (Universidad de San Andres), Chile (Universidad de Chile, Universidad del Norte, Junta de Adelanto de Arica, and the Servicio Nacional de Salud), Ecuador, Peru (Universidad de San Marcos), and the United States (Mayo Clinic, University of Michigan and the University of Texas Health Science Center at Houston). We are indebted to numerous persons and many institutions at the local and national levels in Chile, but we are especially beholden to the Junta de Adelanto de Arica and its president, Luis Beretta. Without the unqualified endorsement of the JAA, and the enthusiastic support of all of the members of the Departamento de Desarollo Comunal y Plan Andino, and particularly Carlos Solari, its chief, this study would not have been possible. His support of this study and concern for the well-being of the Aymara was resolute and unflinching. We also owe much to the Servicio Nacional de Salud, the practicantes of the Carabineros de Chile and numerous teachers who gave selflessly of their time and energies to make our examinations a success. Finally, we express our gratitude to the numerous Aymara who in patience and good humour tolerated our questions, and pokings and proddings.

We gratefully acknowledge the support of the National Institutes of Health through the grant, HL 15614, and the Multinational Genetics Program, Organization of American States, Chile.

References

ALZAMORA-CASTRO, V., ROTTA, A., BATILLANO, G., ABERGATAS, R., RUBIO, C., BOURNOCLE, J., ZAPATA, C., SANTA MARIA, R., BINDER, T., SUBIRIA, R., PANDO, N., PAREDES, D. and PEREZ, V. (1952) Sobre la posible influencia de las grandes alturas en la determinacion de algunas malformaciones cardiacas. *Revista Peruana de Cardiologia Lima* **1**, 259–262.

ARIAS-STELLA, J. and SALDANA, M. (1963) The terminal portion of the pulmonary arterial tree in people native to high altitudes. *Circulation*, **28**, 915–925.

BARCROFT, J. (1923) Observations upon the effect of high altitude on the physiological processes of the human body, carried out in the Peruvian Andes, chiefly at Cerro de Pasco. *Philosophical Transactions of the Royal Society, London* **211,** 351–480.

BENESCH, R. and BENESCH, R. E. (1967) The effect of organic phosphates from the human erythrocyte on the allosteric properties of hemoglobin. *Biochemical and Biophysical Research Communications* **26,** 162–166.

BIRD, J. B. (1943) Excavations in northern Chile. *Anthropological Papers of the American Museum of Natural History*, **38,** 171–316.

BIRD, J. B. (1963) The cultural sequence of the North Chilean Coast. In *Handbook of South American Indians* Vol. II, ed. Steward, J. H., p. 587.

BREWER, G. J. (1974) 2,3-DPG and erythrocyte oxygen affinity. *Annual Review of Medicine* **25,** 29–38.

BUSKIRK, E. (1971) Work and fatigue in high altitude. In *Physiology of Work Capacity and Fatigue* ed. Simonson, E., p. 312. Springfield, Ill.: Charles C. Thomas.

CAEN, J., ERGUETA, J., MICHEL, H., DAUFRESNE, A., POUPART, C. and DHUIME, G. (1971) Modifications de l'agrégation plaquettaire à l'adénosine diphosphate (ADP) chez les Boliviens et les Péruviens de l'Altiplano. *Comptes Rendu de l'Académie des Sciences, Paris,* **272,** 505–508.

CHAKRABORTY, R., SCHULL, W. J., HARBURG, E. and SCHORK, M. A. (1977) Heredity, stress and blood pressure, a family set method. V. Heritability estimates. *Journal of Chronic Disease* (in the press).

CHANUTIN, A. and CURNISH, R. R. (1967) Effect of organic and inorganic phosphates on the oxygen equilibrium of human erythrocytes. *Archives of Biochemistry and Biophysics* **121,** 96–102.

CHEN, S. H., ANDERSON, J. E. and GIBLETT, E. R. (1971) 2,3-diphosphoglycerate mutase: Its demonstration by electrophoresis and the detection of a genetic variant. *Biochemical Genetics* **5,** 481–486.

CRUZ-COKE, R., CRISTOFFANINI, A. P., ASPILLAGA, M. and BIANCANI, F. (1966) Evolutionary forces in human populations in an environmental gradient in Arica, Chile. *Human Biology*, **38,** 421–438.

DEBRUIN, A. H., JANSSEN, L. H. M. and VAN OS, G. A. J. (1971) Effect of 2,3-diphosphoglycerate on the Bohr effect of human adult hemoglobin. *Biochemical and Biophysical Research Communications*, **45,** 544–550.

DONAYRE, J. (1966) Population growth and fertility. In *Life at High Altitudes*, PAHO Publication No. 140, pp. 74–79. Washington D.C.: Pan American Health Organization.

EATON, J. W. and BREWER, G. J. (1968) The relationship between red cell 2,3-diphosphoglycerate and levels of hemoglobin in the human. *Proceedings of the National Academy of Science,* **61,** 756–760.

FORBES, D. (1870) XXIII. On the Aymara Indians of Bolivia and Peru. *Journal of the Ethnological Society of London*, **2,** 192–305.

FRISANCHO, A. R. (1975) Functional adaptation to high altitude hypoxia. *Science,* **187,** 313–319.

HERNANDEZ, Z. (1970) *Geografía de plantas y animales de Chile*, p. 212. Santiago, Chile: Editorial Universitaria.

HURTADO, A. (1964) Acclimatization to high altitudes. In *Physiological Effects of High Altitude*, ed. Weihe, W. H., p. 1. Oxford: Pergamon Press.

HURTADO, A. (1971) The influence of high altitude on physiology. In *High Altitude Physiology: Cardiac and Respiratory Aspects*, eds. Porter, R. and Knight, J., p. 3. Edinburgh and London: Churchill Livingstone.

IATRIDIS, S. G., IATRIDIS, P. G., MARKIDON, S. G. and REGATZ, B. H. (1975) 2,3-di-phosphoglycerate: A physiological inhibitor of platelet aggregation. *Science*, **187**, 257–261.

KELLER, C. (1946) *El Departamento de Arica. Republica de Chile censo economico national*, Vol. 1, p. 344. Santiago, Chile: Zig-Zag.

KEYS, A. (1936) The physiology of life at high altitudes. The international high altitude expedition to Chile, 1935. *Scientific Monthly*, **43**, 289–312.

LABARRE, W. (1948) The Aymara Indians of the Lake Titicaca Plateau, Bolivia. *American Anthropology*, **50**, 1–250.

LENFANT, C., TORRANCE, J., ENGLISH, E., FINCH, C. A., REYNAFARJE, C., RAMOS, J. and FAURA, J. (1968) Effect of altitude on oxygen binding by hemoglobin and on organic phosphate levels. *Journal of Clinical Investigation*, **47**, 2652–2656.

MACNEISH, R. S., BERGER, R. and PROTSCH, R. (1970) Megafauna and man from Ayacucho, highland Peru. *Science*, **168**, 975–977.

MCCLUNG, J. (1969) *Effects of High Altitude on Human Birth*. Cambridge: Harvard University Press.

MERINO, C. F. (1950) Studies on blood formation and destruction in the polycythemia of high altitude. *Blood*, **5**, 1–31.

MONGE, C. and MONGE, C. (1966) *High Altitude Disease: Mechanism and Management*, p. 97. Springfield, Ill.: Charles C. Thomas.

MORPUGO, G., BATTAGLIA, P., BERNINI, L., PAOLUCCI, A. M. and MODIANO, G. (1970) Distribution of HbCO in human erythrocytes following inhalation of CO. *Nature*, **227**, 386–388.

MURRA, J. V. (1968) Aymara kingdom in 1567. *Ethnohistory*, **15**, 115–151.

NIEMEYER FERNANDEZ, H. (1972) *Las pinturas repestres de la sierra de Arica*, p. 114. Santiago, Chile: Editorial Universitaria.

PENALOZA, D., GAMBOA, R., MARTICORENA, E., ECHEVARRIA, M., DYER, J. and GUTIERREZ, E. (1961) The influence of high altitudes on the electrical activity of the heart. *American Heart Journal*, **61**, 101–115.

PENALOZA, D., SIME, F., BANCHERO, N., GAMBOA, R., CRUZ, J. and MARTICORENA, E. (1963) Pulmonary hypertension in healthy men born and living at high altitudes. *American Journal of Cardiology*, **11**, 150–157.

REYNAFARJE, C. (1966) Iron metabolism during and after altitude exposure in man and in adapted animals (camelids). *Federation Proceedings*, **25**, 1240–1242.

REYNAFARJE, C., LORANO, R. and VALDIVIESCO, J. (1959) The polycythemia of high altitudes: Iron metabolism and related aspects. *Blood*, **14**, 433–455.

ROWE, J. H. (1945) Absolute chronology in the Andean area. *American Antiquity*, **10**, 265–284.

SAFFORD, W. E. (1917) Food-plants and textiles of ancient America. *Proceedings of the International Congress of Americanists*, Session XIX, p. 12–30, Washington, 1915.

SANTOLAYA, R., DONOSO, H., APUD, E. and SANUDO, M. C. (1973) Electrocardio-grama y capacidad aerobica en nativos residentes de altura del altiplano chileno, como indice de aclimatación. *Revista Medicina Chileno*, **101**, 433–448.

SCHROTER, W. (1965) Kongenitale nichtspharocytare hämolytische Anämie bei 2,3-Diphosphoglyceratmutase-mangel der Erythrocyten im frühen Saugling-salter. *Klinische Wochenschrift*, **43**, 1147–1153.

SCHULL, W. J. and NEEL, J. V. (1965) *Effects of Inbreeding on Japanese Children*, p. 419. New York: Harper and Row.

SCHULL, W. J., HARBURG, E., ERFURT, J. C., SCHORK, M. A. and RICE, R. (1970) A family set method for estimating heredity and stress. II. Preliminary results of the genetic methodology in a pilot survey of Negro blood pressure, Detroit, 1966–67. *Journal of Chronic Disease*, **23**, 83–98.

SQUIER, E. G. (1968) *Peru: Incidents of Travel and Exploration in the Land of the Incas*, p. 599. London: Macmillan and Co.

STEWART, T. D. (1973) *The Peoples of America*, p. 261. New York: Charles Scribner.

TSCHOPIK, H. (1963) The Aymara. In *Handbook of South American Indians*, Vol. II, ed. Stewart, J. H., p. 501. New York: Cooper Square Publishing Co.

URZUA, L. U. (1969) *Arica, puerta nueva: Historia y folklore*, 3rd edition, p. 291. Santiago, Chile: Editorial Andres Bello.

DE LA VEGA, G. (1604) *Royal Commentaries of the Incas. A General History of Peru*, translated by Livermore, H. V., 1970, p. 1530. Austin, Texas: University of Texas Press.

VIVES-CARRONS, L. L., KAHN, A., HAKIM, J. and BOIVIN, P. (1973) Electrophorèse sur gel d'amidon de la 2,3-diphosphoglycerate mutase erythrocytaire. *Clinica Chimica Acta*, **49**, 91–95.

WEINER, J. S. and LOURIE, J. A. (1969) *Human Biology, a Guide to Field Methods*, p. 621. Oxford: Blackwell Scientific Publications.

WORMALD CRUZ, A. (1968) *Frontera norte*, p. 195. Santiago, Chile: Editorial ORBE.

WORMALD CRUZ, A. (1969) *El mestizo en el Departamento de Arica*, p. 210. Santiago, Chile: Ediciones Rafaga.

AUTHOR INDEX

Page numbers in parentheses are locations of full references for papers cited in the text.

171

SUBJECT INDEX

177

SOCIETY FOR THE STUDY OF HUMAN BIOLOGY

Although there are many scientific societies for the furtherance of the biological study of man as an individual, there has been no organization in Great Britain catering for those (such as physical anthropologists or human geneticists) concerned with the biology of human populations. The need for such an association was made clear at a Symposium at the Ciba Foundation in November 1957, on "The Scope of Physical Anthropology and Human Population Biology and their Place in Academic Studies". As a result the Society for the Study of Human Biology was founded on May 7th, 1958, at a meeting at the British Museum (Natural History).

The aims of the Society are to advance the study of the biology of human populations and of man as a species, in all its branches, particularly human variability, human genetics and evolution, human adaptability and ecology.

At present the Society holds two full-day meetings per year—a Symposium (usually in the autumn) on a particular theme with invited speakers, and a scientific meeting for proffered papers. The papers given at the Symposia are published and the monographs are available to members at reduced prices.

Persons are eligible for membership who work or who have worked in the field of human biology as defined in the aims of the Society. They must be proposed and seconded by members of the Society. The subscription is £5.50 per annum (this includes the Society's journal *Annals of Human Biology*) and there is no entrance fee.

Applications for membership should be made to Dr. A. J. Boyce, Hon. General Secretary, Department of Biological Anthropology, 58 Banbury Road, Oxford OX2 6QS.

PUBLICATIONS OF THE SOCIETY

Symposia, Volume V, 1963: *Dental Anthropology*, edited by D. R. BROTH-WELL. Pergamon Press (members £1.25).

Symposia, Volume VI, 1964: *Teaching and Research in Human Biology*, edited by G. A. HARRISON. Pergamon Press (members £1.25).

Symposia, Volume VII, 1965: *Human Body Composition, Approaches and Applications*, edited by J. BROZEK. Pergamon Press (members £3).

Symposia, Volume VIII, 1968: *The Skeletal Biology of Earlier Human Populations*, edited by D. R. BROTHWELL. Pergamon Press (members £2).

Symposia, Volume X, 1971: *Biological Aspects of Demography*, edited by W. BRASS. Taylor & Francis (members £2.50).

Symposia, Volume XI, 1973: *Human Evolution*, edited by M. H. DAY. Taylor & Francis (members £2.50).

Symposia, Volume XII, 1973: *Genetic Variation in Britain*, edited by D. F. ROBERTS and E. SUNDERLAND. Taylor & Francis (members £3.50).

Symposia, Volume XIII, 1975: *Human Variation and Natural Selection*, edited by D. F. ROBERTS. Taylor & Francis (members £2.75).

Symposia, Volume XIV, 1975: *Chromosome Variation in Human Evolution*, edited by A. J. BOYCE. Taylor & Francis (members £3.00).

Symposia, Volume XV, 1976: *The Biology of Human Fetal Growth*, edited by D. F. ROBERTS and A. M. THOMSON. Taylor & Francis (members £3.75).

Symposia, Volume XVI, 1977: *Human Ecology in the Tropics*, edited by J. P. GARLICK and R. W. J. KEAY. Taylor & Francis (members £3.00).

Symposia, Volume XVII, 1977: *Physiological Variation and its Genetic Basis*, edited by J. S. WEINER. Taylor & Francis (members £4.00).